Antarctic Sea Ice Variability in the Southern Ocean-Climate System

Proceedings of a Workshop

Alison Macalady and Katie Thomas, *Rapporteurs*

Polar Research Board

Ocean Studies Board

Division on Earth and Life Studies

The National Academies of
SCIENCES · ENGINEERING · MEDICINE

THE NATIONAL ACADEMIES PRESS
Washington, DC
www.nap.edu

THE NATIONAL ACADEMIES PRESS 500 Fifth Street, NW Washington, DC 20001

This project was supported by the National Aeronautics and Space Administration (Award No. NNX15AP72G) and the National Science Foundation (Award No. PLR-1540067). Any opinions, findings, conclusions, or recommendations expressed in this publication do not necessarily reflect the views of any organization or agency that provided support for the project.

International Standard Book Number-13: 978-0-309-45600-5
International Standard Book Number-10: 0-309-45600-2
Digital Object Identifier: 10.17226/24696

Additional copies of this publication are available for sale from the National Academies Press, 500 Fifth Street, NW, Keck 360, Washington, DC 20001; (800) 624-6242 or (202) 334-3313; http://www.nap.edu.

Printed in the United States of America

Cover credit: Edge of an ice shelf in Adelaide Island, off the Antarctic Peninsula. NASA/Maria-Jose Vinas.

Suggested citation: National Academies of Sciences, Engineering, and Medicine. 2017. *Antarctic Sea Ice Variability in the Southern Ocean-Climate System*. Washington, DC: The National Academies Press. doi: 10.17226/24696.

The National Academies of

SCIENCES · ENGINEERING · MEDICINE

The **National Academy of Sciences** was established in 1863 by an Act of Congress, signed by President Lincoln, as a private, nongovernmental institution to advise the nation on issues related to science and technology. Members are elected by their peers for outstanding contributions to research. Dr. Marcia McNutt is president.

The **National Academy of Engineering** was established in 1964 under the charter of the National Academy of Sciences to bring the practices of engineering to advising the nation. Members are elected by their peers for extraordinary contributions to engineering. Dr. C. D. Mote, Jr., is president.

The **National Academy of Medicine** (formerly the Institute of Medicine) was established in 1970 under the charter of the National Academy of Sciences to advise the nation on medical and health issues. Members are elected by their peers for distinguished contributions to medicine and health. Dr. Victor J. Dzau is president.

The three Academies work together as the **National Academies of Sciences, Engineering, and Medicine** to provide independent, objective analysis and advice to the nation and conduct other activities to solve complex problems and inform public policy decisions. The National Academies also encourage education and research, recognize outstanding contributions to knowledge, and increase public understanding in matters of science, engineering, and medicine.

Learn more about the National Academies of Sciences, Engineering, and Medicine at **www.national-academies.org**.

The National Academies of

SCIENCES · ENGINEERING · MEDICINE

Reports document the evidence-based consensus of an authoring committee of experts. Reports typically include findings, conclusions, and recommendations based on information gathered by the committee and committee deliberations. Reports are peer reviewed and are approved by the National Academies of Sciences, Engineering, and Medicine.

Proceedings chronicle the presentations and discussions at a workshop, symposium, or other convening event. The statements and opinions contained in proceedings are those of the participants and have not been endorsed by other participants, the planning committee, or the National Academies of Sciences, Engineering, and Medicine.

For information about other products and activities of the National Academies, please visit nationalacademies.org/whatwedo.

PLANNING COMMITTEE ON ANTARCTIC SEA ICE VARIABILITY IN THE SOUTHERN CLIMATE-OCEAN SYSTEM

JULIENNE STROEVE (*Chair*), University of Colorado Boulder
DAVID HOLLAND, New York University, New York
MARIKA HOLLAND, National Center for Atmospheric Research, Boulder, Colorado
TED MAKSYM, Woods Hole Oceanographic Institution, Woods Hole, Massachusetts
MARILYN RAPHAEL, University of California, Los Angeles
SUSAN SOLOMON, Massachusetts Institute of Technology, Cambridge
XIAOJUN YUAN, Columbia University, Palisades, New York

National Academies of Sciences, Engineering, and Medicine Staff
KATIE THOMAS, Senior Program Officer, Polar Research Board
ALISON MACALADY, Program Officer, Polar Research Board (until August 2016)
AMANDA STAUDT, Director, Polar Research Board
YASMIN ROMITTI, Research Associate, Polar Research Board
ROB GREENWAY, Program Associate, Polar Research Board
MICHAEL HUDSON, Senior Program Assistant, Polar Research Board
SHELLY FREELAND, Financial Associate, Polar Research Board

Acknowledgments

This Proceedings of a Workshop has been reviewed in draft form by individuals chosen for their diverse perspectives and technical expertise. The purpose of this independent review is to provide candid and critical comments that will assist the institution in making its published Proceedings of a Workshop as sound as possible and to ensure that the Proceedings of a Workshop meets institutional standards for objectivity, evidence, and responsiveness to the study charge. The review comments and draft manuscript remain confidential to protect the integrity of the deliberative process. We wish to thank the following individuals for their review of this Proceedings of a Workshop:

Ian Allison, University of Tasmania, Australia
Elizabeth Hunke, Los Alamos National Laboratory, Santa Fe, NM
Claire Parkinson, NASA Goddard Space Flight Center, Greenbelt, MD
Marilyn Raphael, University of California, Los Angeles

Although the reviewers listed above have provided many constructive comments and suggestions, they did not see the final draft of the Proceedings of a Workshop before its release. The review of this Proceedings of a Workshop was overseen by **William Berry Lyons**, The Ohio State University, who was responsible for making certain that an independent examination of this Proceedings of a Workshop was carried out in accordance with institutional procedures and that all review comments were carefully considered. Responsibility for the final content of this Proceedings of a Workshop rests entirely with the rapporteurs and the institution.

The planning committee would like to thank the staff at the Cooperative Institute for Research in Environmental Sciences at the University of Colorado Boulder for graciously hosting the workshop. Special thanks are also due to those who gave plenary presentations, were note-takers, or were rapporteurs for the breakout groups.

Contents

Antarctic Sea Ice Variability in the Southern Ocean-Climate System: Proceedings of a Workshop

OVERVIEW

The sea ice surrounding Antarctica has increased in extent and concentration from the late 1970s, when satellite-based measurements began, until 2015.[1] Although this increasing trend is modest, it is surprising given the overall warming of the global climate and the region. Indeed, climate models, which incorporate our best understanding of the processes affecting the region, generally simulate a decrease in sea ice. Moreover, sea ice in the Arctic has exhibited pronounced declines over the same period, consistent with global climate model simulations. For these reasons, the behavior of Antarctic sea ice has presented a conundrum for global climate change science.

The National Academies of Sciences, Engineering, and Medicine held a workshop on January 11-12, 2016, in Boulder, Colorado, to bring together scientists with different sets of expertise and perspectives to further explore potential mechanisms driving the evolution of recent Antarctic sea ice variability[2] and to discuss ways to advance understanding of Antarctic sea ice and its relationship to the broader ocean-climate system.

The significant distinctions between the Southern and Arctic Oceans, which are different dynamic systems, was a common theme of the workshop discussions. In particular, the Southern Ocean surface is relatively insensitive to recent human-caused greenhouse gas (GHG) forcing, so surface waters have experienced less warming than has been observed in other areas. However, changes in the deep ocean can have an outsized effect on the Southern Ocean because it is a region of significant upwelling. Therefore, it is expected that as the global deep water continues to warm, it will continue to melt Antarctic ice for many decades, even if carbon dioxide (CO_2) levels are reduced.

There are many local, regional, and global processes that influence sea ice growth and melt, but it is not clear what mechanisms best explain the observed variability and the slight increase in overall Antarctic sea ice extent. Some participants highlighted how negative feedbacks between the ocean and sea ice keep Antarctic sea ice thin. Other studies suggest that there may be a modest decrease in Antarctic sea ice extent from anthropogenic warming, but that the trend is overwhelmed by a recent increase associated with high internal variability, resulting in the recent slight expansion of Antarctic sea ice extent. Other factors that may play a role include tropical Pacific and Atlantic teleconnections, variability in the winds and ocean currents that

---

[1] Note that Antarctic sea ice extent has experienced significant decreases since the workshop was held in January 2016. See http://nsidc.org/data/seaice_index.

[2] In the context of this workshop, sea ice variability primarily refers to variability in sea ice extent, but it also includes variability in sea ice concentration, area, velocity, thickness, and snow cover.

circumnavigate Antarctica, and changes in surface conditions resulting from stratospheric ozone depletion.

Many workshop discussions emphasized the distinct regional variability of Antarctic sea ice patterns. For example, most of the increase in total sea ice extent has actually occurred in the western Ross Sea. The drivers of the larger increases in this region are not well understood, and the observed trend is outside the range of forced response and internal variability. Some participants said that variability in tropical sea surface temperature (SST) may be leading to changes in winds that have in turn caused cooling in the Ross Sea. This is a key region for improved observations and process studies to better understand the relationship between polynyas, ice production topography, and ice shelf melt processes. In contrast, the sea ice cover and duration in the Bellingshausen Sea are decreasing. Some studies indicate that changes in the Amundsen Sea Low (ASL) and increases in ocean heat content are driving this decrease.

Understanding the mechanisms and processes driving sea ice variability and trends is limited by the lack of a long, homogenous record of sea ice extent and concentration. Furthermore, observations of the surrounding ocean are geographically sparse and short. According to several participants, more effort should go into extending the observational record using proxies, historical records, and data assimilation. At this time, the data that have been captured to extend the historical record indicate a larger sea ice extent prior to the satellite period. However, questions were raised as to whether there is enough confidence in the diverse set of proxies to make such a definitive statement. In addition, some participants highlighted the need for more validation of the passive microwave observations of sea ice concentrations taken from space-based platforms. Reanalysis data also remain problematic and it is not clear which reanalysis is best suited for Antarctic studies.

Some participants said that there is little confidence in the models that are used for attribution of Antarctic sea ice variability. For example, many of the models have a poor representation of the mean state of the Southern Ocean, which is important to reproduce observed trends in Antarctic sea ice. Furthermore, model biases (e.g., stratification/mixed layer depth) impact how the models respond to forcing, but there is uncertainty on how large an impact the biases have. Despite these shortcomings, said many participants, models are useful for understanding processes that may be affecting sea ice.

Many participants said that process-based understanding is critical for improving understanding of the mechanisms of Antarctic sea ice variability. Process studies also provide validation of global coupled models. However, higher-resolution atmosphere and ocean models may be necessary to more fully understand some important processes, especially for resolving eddies, mixed layer depth, polynyas/ice formation, and katabatic winds/cyclogenesis. Another barrier to process-based understanding is the dearth of observations in the Southern Ocean (e.g., under the sea ice), particularly those that are necessary to make heat and freshwater budget calculations. Such calculations require estimates of sea ice thickness; estimates of precipitation over the Southern Ocean; exchange of water between the ocean and under-ice-shelf cavities; and hydrographic measurements of temperature, salinity, and isotopes.

Models that project climate conditions decades and longer into the future indicate that the Antarctic sea ice will eventually respond to global warming and decline. Observations from late

2016 and early 2017[3] indeed show decreases in Antarctic sea ice extent. Nevertheless, many participants said a better understanding of the mechanisms is critical to making confident statements about the future of Antarctic sea ice.

INTRODUCTION

Sea ice in the oceans surrounding Antarctica has been more extensive on average in the few years prior to the workshop (2012-2014) than at any other time in the satellite-observed record. There is not a consensus on the causes of Antarctic sea ice trends or regional variability, and observed patterns are not well represented in Earth system models. Given the importance of sea ice in the Earth system (see Box 1) and gaps in understanding of its past variability and likely future fate in the Southern Hemisphere, the Polar Research Board and the Ocean Studies Board of the National Academies of Sciences, Engineering, and Medicine organized a workshop to evaluate recent sea ice variability and change in the Southern Ocean surrounding Antarctica (the full statement of task can be found in Appendix A). Specific purposes of the workshop included

- Examining the observational and modeled record of recent (last 50 years) Antarctic sea ice variability,
- Evaluating evidence for key hypotheses about the processes controlling recent Antarctic sea ice variability and change, and
- Highlighting knowledge gaps and important areas for new research that could clarify the mechanisms of past sea ice variability and help constrain projections of future Antarctic sea ice variability.

The workshop was planned and moderated by a seven-member ad hoc committee appointed by the National Academies (the committee member biographical sketches are in Appendix F) and was held on January 11-12, 2016, in Boulder, Colorado. The event brought together domestic and international experts in remotely sensed and in situ sea ice observations, as well as scientists who study sea ice dynamics, oceanography, climatology, and sea ice and climate modeling (see Appendix D for a list of participants).

The committee organized the workshop roughly around the project's statement of task (see Appendix A), with sessions on observations of Antarctic sea ice and the surrounding ocean and atmosphere, model projections of Antarctic sea ice variability, and the attribution of recent sea ice variability and change to various processes in the coupled ocean-atmosphere-ice system. An introductory session included longer, summary talks that provided an overview and a critical entrée into each of these three broad topics. Participants also broke into smaller groups to analyze the session topics in more detail. The agenda for the workshop can be found in Appendix D.

The organization of this proceedings also roughly follows the statement of task and presents topics thematically rather than following the chronology of the workshop. The first section provides an overview of the observational record Antarctic sea ice variability, including

[3] See http://nsidc.org/data/seaice_index.

BOX 1
Why Is Understanding Antarctic Sea Ice Variability Important?

Sea ice plays a critical role in the climate system, regulating the flux of heat, gases, and momentum between the atmosphere and the ocean. A recent attempt to quantify the overall impact of sea ice on the current climate found that sea ice and anthropogenic greenhouse gas emissions are of similar magnitude in terms of their influence on the global heat budget (Cvijanovic and Caldiera, 2015). The sea ice–albedo feedback—in which melting sea ice can accelerate climate warming by exposing dark ocean areas that absorb more sunlight—is widely recognized as an important reason for recent amplification of climate warming in the Arctic (Screen and Simmonds, 2010). Changes in the Antarctic, where average sea ice extent is approximately 20% greater than in the Arctic, could result in relatively large changes to planetary albedo. Antarctic sea ice further influences the climate system by regulating heat and carbon dioxide export from the atmosphere to the ocean. The Southern Ocean currently absorbs about one-sixth of annual human emissions of CO_2 and is responsible for a disproportionate amount of the global ocean uptake of heat (Frölicher et al., 2015). Sea ice and associated polynyas play an important role modulating the transfer of this heat from the atmosphere to Southern Ocean deep waters through their influence on the salinity and density of surface waters (Bitz et al., 2006; Kirkman and Bitz, 2011).

Furthermore, feedbacks between sea ice production and ocean water temperature and salinity may play a role in determining the stability of Antarctica's massive sheets of glacial ice (Khazendar et al., 2013; Massom et al., 2010; Miles et al., 2016). Understanding sea ice variability and trends may thus be important for anticipating the rate of ice sheet melt and sea level rise in the coming decades.

Third, Antarctica's ecosystems are critically dependent on sea ice. Sea ice is a leading determinant of marine ecosystem structure and function, affecting nutrient dynamics, light availability, and ocean salinity and temperature gradients (Massom and Stammerjohn, 2010). A large portion of the productivity in the Southern Ocean happens within the sea ice, not in the open water (Arrigo et al., 2009), and algal communities within the ice provide a critical food source for krill year round. These same communities also contribute to massive phytoplankton blooms during the spring and summer melt season. Any changes to sea ice–dependent communities at the bottom of the food chain affect fish and other animals higher on the trophic ladder (Massom and Stammerjohn, 2010).

observed differences in the regions and seasons (see Figure 1). This is followed by a section on the use of models to better understand the drivers of Antarctic sea ice. The next section provides more detail on a range of proposed processes and mechanisms that control observed sea ice variability and change. The final section summarizes individual workshop participants' views on future needs and opportunities associated with observing, modeling, and understanding Antarctic sea ice variability.

This Proceedings of a Workshop was prepared by workshop rapporteurs as a factual summary of the workshop presentations and discussions. The planning committee's role was limited to planning and convening the workshop. The views contained in this proceedings are those of individual workshop participants and do not necessarily represent the views of all workshop participants, the planning committee, or the National Academies of Sciences, Engineering, and Medicine. Thus, this Proceedings of a Workshop is not intended to provide an exhaustive overview of the rapidly progressing field of research on Antarctic sea ice, but rather to provide a current and concise window into critical issues by summarizing the discussions of leading U.S. and international scientists.

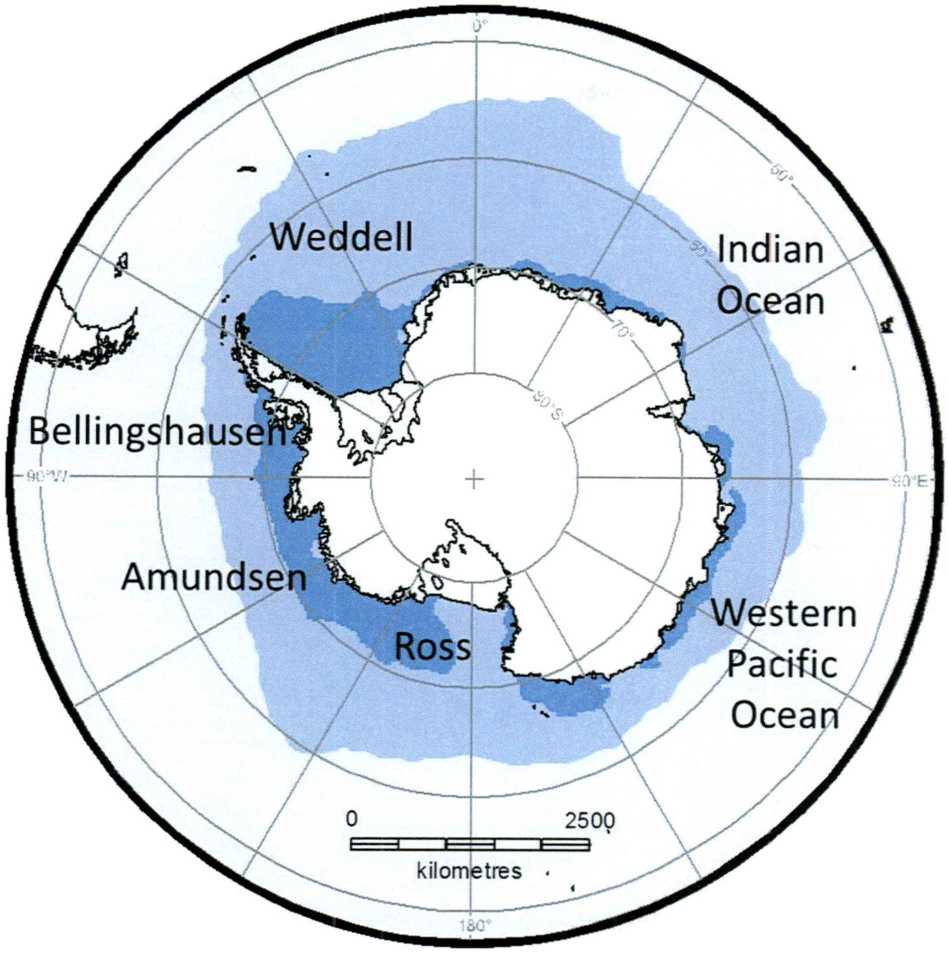

FIGURE 1 Map of Antarctica and different region sectors. Light blue shading indicates mean winter sea ice extent; dark blue shading indicates mean summer sea ice extent. SOURCE: Presented by Dr. Sarah Das, figure modified from Abram et al., 2013.

OBSERVATIONAL RECORD OF ANTARCTIC SEA ICE VARIABILITY

Background

The Antarctic experienced modest increases in maximum and minimum sea ice extent and sea ice concentration since late 1978, when routine satellite measurements started, until 2014 (Comiso et al., 2011; Meier et al., 2013; Parkinson and Cavalieri, 2012; Turner et al., 2015; Zwally et al., 2002). The record maximum extents have all occurred in recent years, with 2012-2014 progressively setting records for September monthly sea ice extent (Massonnet et al., 2015; Parkinson, 2014; Reid et al., 2015; Turner et al., 2013a). The sea ice returned to near-average extent in 2015 (Fetterer et al., 2016) and was below average in 2016 (Fetterer et al., 2016; see Figure 2).

The overall increase in Antarctic sea ice extent stands in contrast to conditions in the Arctic, where sea ice has exhibited pronounced declines (the average monthly sea ice index has been declining at a rate of –2.7% per decade for March, the month in which sea ice reaches its maximum extent in the Northern Hemisphere), and a series of recent record minimum extents (Parkinson and Comiso, 2013). The direction of the Arctic trend is consistent with global climate

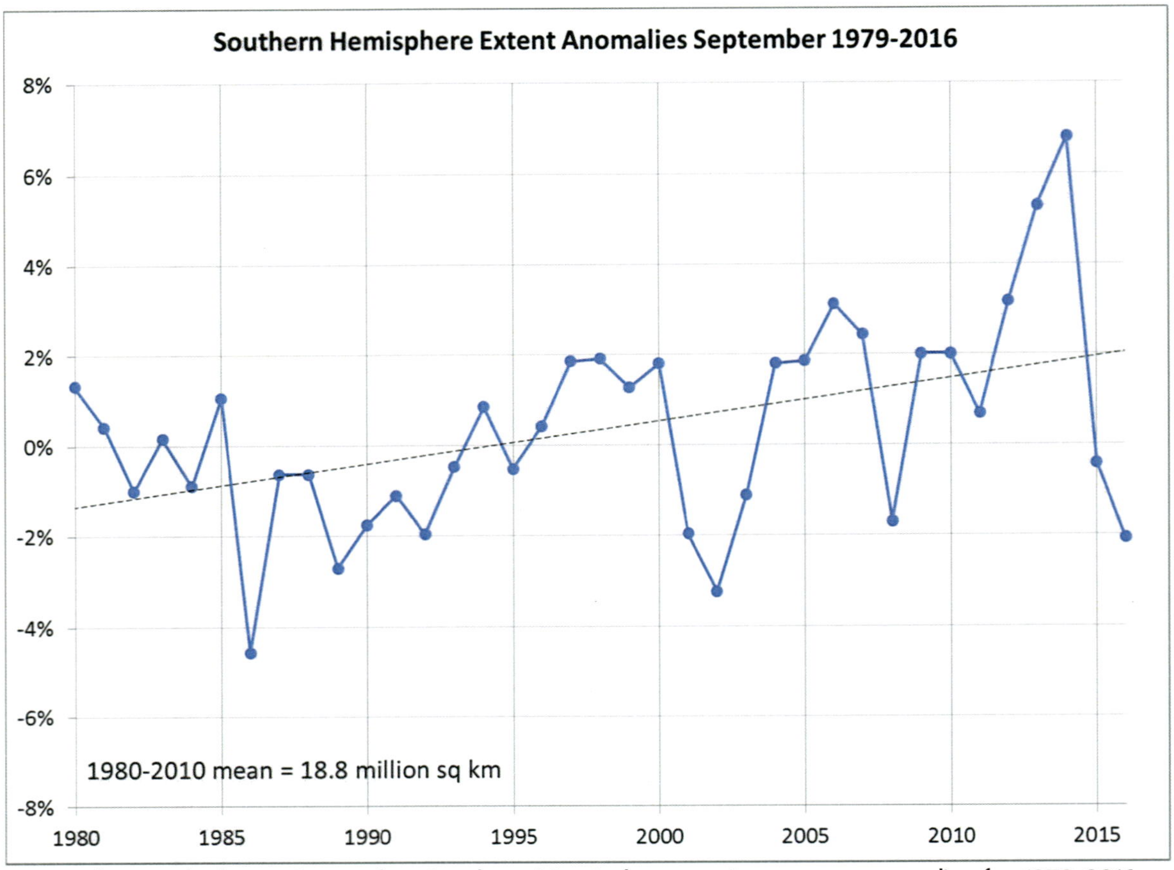

FIGURE 2 This graph shows September Southern Hemisphere sea ice extent anomalies for 1979-2016, plotted as a time series of percent difference between the extent for each September and the mean September extent over the 30-year period 1980-2010. The anomaly data points are plotted as circles and the trend line is plotted with a dashed grey line. The overall trend is slightly positive compared to the 30-year mean (1981-2010). SOURCE: http://nsidc.org/data/seaice_index.

model simulations, and shrinking Arctic sea ice is frequently cited as a prominent indicator of greenhouse gas–induced global warming (Stroeve et al., 2012).

The observed Antarctic sea ice record is often presented as surprising and as a conundrum for global climate change science (e.g., Massonnet et al., 2015; Turner et al., 2015). Antarctic sea ice formation and persistence are sensitive to air and sea surface temperatures, and considering that Antarctic air temperatures and Southern Ocean SSTs have increased slightly since 1950 (Armour and Bitz, 2015; Turner et al., 2015), the trends in Antarctic sea ice are not necessarily intuitive (e.g., Turner et al., 2015). However, SST trends have not been as strong in the Southern Ocean when compared to other parts of the Earth (Armour et al., 2016; Turner et al., 2015), and sea ice extent and thickness are variably influenced by other thermodynamic and dynamic factors such as transport by surface winds (Holland and Kwok, 2012; Maksym et al., 2012), snowfall and rain rates above the ice surface (Massom et al., 2001), and water temperatures and salinity beneath the sea surface (Martinson and Iannuzzi, 1998).

The Satellite Record

Observations of sea ice extent, concentration, and area derived from passive microwave radiometers constitute the longest-running and most geographically extensive of sea ice records, and these observations are almost universally drawn upon for studies of sea ice variability and change in both hemispheres, according to Dr. Julienne Stroeve, National Snow and Ice Data Center, and Dr. Claire Parkinson, NASA Goddard Space Flight Center. Routine satellite observation of continental-scale sea ice variability began in the late 1970s with the launch of satellites carrying passive microwave radiometers. Data from such radiometers can be used to estimate the amount of ice coverage in a given area, which is usually expressed in terms of sea ice extent, area, or concentration.

Dr. Rob Massom, Australian Antarctic Division, outlined key uncertainties in the satellite observational record. He noted that the accuracy and precision of passive-microwave-derived sea ice concentration and extent retrievals are affected by a combination of data, algorithm, and physical (environmental) factors. While mismatches in calibration, resolution, and scanning geometry between satellite sensors can lead to spurious retrievals, retrieval uncertainties can be minimized by careful reprocessing (Comiso and Nishio, 2008; Meier et al., 2011). Indeed, Dr. Massom said that in previous reports the significance of the observed trend from the passive microwave record may be compromised by limitations in retrieval algorithm accuracy and precision (cf., Eisenman et al., 2014; Meier et al., 2011) and are no longer valid, as the issues were rectified prior to Intergovernmental Panel on Climate Change Assessment Report 5 (IPCC, 2013).

According to Drs. Parkinson, Stroeve, and Massom, although there is still important work to be done improving the algorithms that transform raw sensor radiative data into time series of sea ice extent, area, and concentration, the magnitude of variability and the overall trends in sea ice remain very similar regardless of the algorithm used (e.g., IPCC, 2013; Parkinson and Comiso, 2008). Absolute differences in retrievals from the different major algorithms (e.g., NASA Team and Bootstrap) are caused by differences in the choice and utilization of channels, tie-points used, sensitivities to changes in surface physical temperature and emissivity, and use or otherwise of weather filters (Comiso et al., 1997, 2003). As a general rule of thumb, said Dr. Massom, overall accuracies in ice concentration retrieval ranging from ±5% to ±15% are expected. Errors are largest in summer, over regions of predominantly thin/new ice, and year round in the marginal ice zone (MIZ) (up to ±30%) (Brucker et al., 2014; Comiso et al., 2011). Further calibration and validation work is needed to address these uncertainties and to quantify satellite algorithm and data accuracy/bias and precision as a function of region and season (Brucker et al., 2014).

The highly dynamic Antarctic sea ice edge region/MIZ is a particular challenge to observe with passive microwave sensors due to the predominance of new/thin ice and ice surface wetness (caused by wave-ice interaction) and strong weather effects, said Dr. Massom. Sea ice concentration retrieval errors increase within the MIZ due in large part to the overlapping microwave emissivities of wet/thin sea ice and open ocean. There is also a melt-season bias: passive microwave observations of ice edge location are typically in reasonable agreement with in situ observations during the austral cold season (March-October), but not in the melt season (November-February). At this time, a 1-1.5 degree latitudinal discrepancy has been noted, with the satellite-derived ice edge being south of the actual edge (Worby and Comiso, 2004). Dr. Massom also noted that "ice edge" is a somewhat ambiguous term, given that the outer sea ice zone around Antarctica is typically indistinct and characterized by diffuse low-concentration bands (depending on wind direction). For this reason, sea ice edge location is typically defined by the outer 15% ice-concentration isoline.

Dr. Massom outlined work that has been done to use satellite data from before the modern multifrequency passive microwave era (i.e., before late 1978) to extend the observational record back in time. Cavalieri et al. (2003) reprocessed sea ice extents from the single-channel Nimbus[4] 5 Electrically Scanning Microwave Rediometer (ESMR) (1973-1976). Including these data changes the trend in overall net Antarctic sea ice extent to negative, as there is evidence that sea ice extent was larger in 1973-1975. However, there is unknown bias in the ESMR data, so they are not often used in the calculation of trends (IPCC, 2013).

Data mining efforts have been ongoing to derive snapshots of Antarctic sea ice extent from early Nimbus satellite visible and infrared imagery from the 1960s. Results from these efforts indicate

- Higher overall extents in the late 1960s (Sissala et al., 1972; Streten, 1973; Zwally et al., 1983) compared to the post-1979 record;
- A September 1964 sea ice extent (Meier et al., 2013) that was higher than any annual maximum in the passive microwave time series of 1979-2013, including records in daily maxima in 2012 and 2013, respectively (Reid et al., 2015); and
- Antarctic monthly sea ice extents in May-July 1966 that were similar to those from the same time of year in the 1980s (Gallaher et al., 2014).

According to Dr. Massom, these early data may be unreliable due to uneven sampling, large uncertainties, and unknown biases. Navigation and cloud-ice discrimination errors may introduce a bias compared to passive microwave estimates, so it is difficult to quantify and compare these early data to the data from microwave sensors. This is also affected by the difference in spatial resolution of the two sea ice extent products. However, said Dr. Massom, such data mining exercises are key to placing the post-1979 record into longer-term context. Considerable decadal variability in observed sea ice extent trends in the satellite passive microwave record to date underline the crucial need to extend the time series (e.g., Simpkins et al., 2013).

While there is a tendency to focus on overall sea ice extent and concentration, these are incomplete descriptors of the sea ice "state," according to Dr. Massom. In addition, there is a crucial need for additional information on sea ice thickness and volume (IPCC, 2013). Lower sea ice extent does not necessarily equate to thinner sea or lower ice volume. For example, divers measured sea ice thicknesses of 20 m in the Bellingshausen Sea in October 2001 that resulted from extreme ice compaction against the Antarctic Peninsula driven by strong and persistent northwesterly winds (Massom et al., 2006). However, this coincided with a period of anomalously low regional ice extent. This example underlines the importance of improved information on sea ice motion and dynamics as they affect sea ice extent and thickness, and it also speaks to the importance of extreme atmospheric events in driving regional changes in sea ice extent and

---

[4] See http://nssdc.gsfc.nasa.gov/earth/nimbus.html.

thickness. Also crucial is improved information on precipitation rates and snow thickness as they affect sea ice formation and melt (Leonard and Maksym, 2011; Maksym and Markus, 2008).

Coastal landfast sea ice (fast ice) forms a narrow yet important interface between ice sheet and ocean/pack ice and is also a sensitive indicator of change and variability in wind and ocean wave fields. To date, fast ice has been mapped at 2 km resolution around the East Antarctic coast only using MODIS visible/infrared imagery. This time series (from 2000 to 2008) shows an upturn in coastal fast ice overall, particularly after 2004 (Fraser et al., 2012), but with strong regionality. Fast ice extent in the western Pacific Ocean sector shows significant interannual variability, while that in the southeastern Indian Ocean sector showed an upturn in 2004, largely due to increase in summer persistence. Dr. Massom acknowledged the shortness of this time series and noted that extending the record in time and to circumpolar is in progress.

Dr. Nathan Kurtz, NASA Goddard Space Flight Center, discussed the contribution of Antarctic sea ice thickness observations to understanding of sea ice volume change. Comparisons of 2003-2008 sea ice thickness observations from ICESat (a laser altimeter; Kurtz and Markus, 2012) and 1981-2005 ship-based observations (Worby et al., 2008) show no significant differences. This suggests that there have not been significant changes in the ice thickness over the two time periods, according to Dr. Kurtz.

Dr. Kurtz said radar altimetry (ERS-1, ERS-2, Envisat, CryoSat-2[5]) offers the longest time series (since 1991), but a number of challenges need to be addressed to get accurate retrievals of sea ice thickness. One challenge is that over Antarctic sea ice, the radar does not penetrate to the snow/ice interface but to somewhere between the air/snow and snow/ice interfaces (Giles et al., 2008). Another challenge is that only small changes have been observed in Antarctic sea ice thickness and volume, so longer time series and/or more accurate retrievals are needed in order to determine statistically significant trends, according to Dr. Kurtz.

Dr. Kurtz identified several major sources of uncertainty from altimeters:

- Laser: open water identification, clouds/forward scattering, zero freeboard assumption, snow depth/density, and sea ice density.
- Radar: waveform tracking, absorption/penetration, backscatter from snow and ice layers, off-nadir returns from smooth surfaces, and snow depth on sea ice.

Radar altimetry should be combined with other sources of data (e.g., sea ice thickness measurements from ships) to improve the sea ice thickness record, to quantify spatiotemporal variability of snow and sea ice density, and to validate the satellite data, said Dr. Kutz. Examples of datasets that could be combined with satellite date include in situ measurements, undersea sonar, autonomous underwater vehicles, electromagnetic (EM) bird (airborne electromagnetic instruments), and the passive microwave record. Operation IceBridge[6] is providing a novel combination of laser and radar measurements, but there are questions on how best to use the limited flight lines.

---

[5] See http://www.esa.int/Our_Activities/Observing_the_Earth/CryoSat/Introducing_CryoSat.
[6] IceBridge, a 6-year NASA mission, is a large airborne survey of Arctic and Antarctic ice sheets, ice shelves, and sea ice. For more information, see https://www.nasa.gov/mission_pages/icebridge/index.html.

Dr. Ron Kwok, Jet Propulsion Laboratory, California Institute of Technology, noted that the new NASA mission, IceSat-2, will calculate sea ice freeboard and sea surface height.

In Situ Observations

Dr. Michael Meredith, British Antarctic Survey, said that ocean observations are critical for understanding Antarctic sea ice, especially given that the ocean is an integral component of the high-latitude air-sea-ice system (see Box 2 which describes the role of the Southern Ocean in Antarctic sea ice). SST from satellite observational trends shows a relationship to sea ice trends (Armour and Bitz, 2015), but a genuine mechanistic understanding requires more than just surface data.

The Southern Ocean is the key site globally where deep waters are returned to the surface and surface waters are downwelled (see Figure 3). But the subsurface Southern Ocean (especially in the sea ice zone) is one of the worst measured places in the world, said Dr. Meredith. The Southern Ocean in winter is virtually unsampled, aside from Argo,[7] satellites, and a handful of moorings (see Figure 4).

More sustained observations are needed, said Dr. Meredith, but they need to be planned and coordinated strategically, so that the right data are collected in the right place at the right time, in order to tackle pressing scientific and societal problems. Sea ice matters, but it is one part of a very complex and undermeasured system (Meredith et al., 2013).

Dr. Katherine Leonard, University of Colorado Boulder, said in situ precipitation measurements for the Southern Ocean sea ice are available, and there have been a variety of attempts at collecting these measurements since the 1960s. However, these measurements are challenging because most precipitation is either blown off into the ocean or accumulates in ridges (Leonard and Cullather, 2008).

BOX 2
The Role of the Ocean

Overall, the surface Southern Ocean has warmed more slowly than the global ocean since 1950. Dr. Kyle Armour, University of Washington, noted that south of the Antarctic Circumpolar Current (ACC), the SST has warmed only around 25% as much as elsewhere around the global ocean. Models are able to replicate less warming in the Antarctic versus Arctic very well. The mechanism that is traditionally proposed for this slow warming is efficient deep ocean heat uptake due to deep mixed layers and enhanced upper ocean stratification, leading to reduced upward heat flux (e.g., IPCC, 2013; Manabe et al., 1991). In the Southern Ocean there is upwelling of cooler, deeper waters that do not have time to warm at the surface. There are also divergent flows; a stronger northward flow that transports heat away from the continent and a smaller southward flow that brings heat toward the continent. This is in stark contrast to the Arctic, where there is no comparable upwelling mechanism and the Arctic Ocean is therefore free to warm quickly under climate forcing; sea ice changes in the Southern Ocean and Arctic Ocean should not be directly compared to one another (Armour et al., 2016; Marshall et al., 2014).

---

[7] Argo is a major component of the ocean observing system for observing temperature, salinity, and currents in Earth's oceans. For more information, see http://www.argo.ucsd.edu/About_Argo.html.

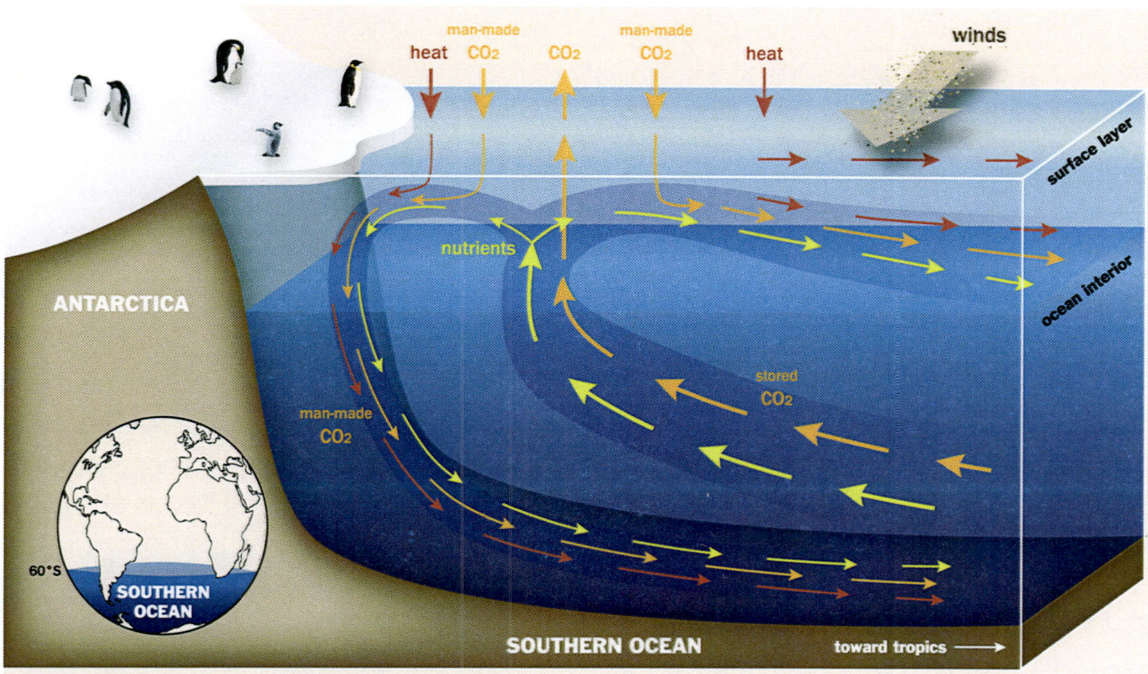

FIGURE 3 Schematic depicting Southern Ocean circulation, which is characterized by deep water returning to the surface where it is reprocessed into new water masses. SOURCE: Figure by Ilissa Ocko, courtesy of Princeton University, presented by Dr. Michael Meredith.

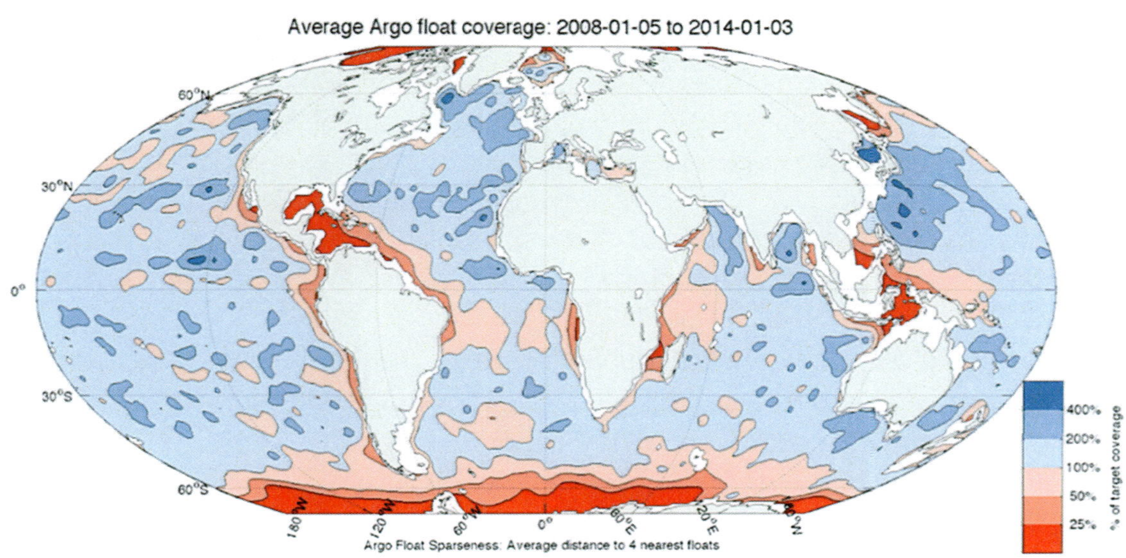

FIGURE 4 Argo coverage (density of number of floats per degree square) from 2008 to 2014 in the world ocean. Much of the Southern Ocean sea ice zone has only 25% of target coverage (red). The target total is 3,200 floats globally. SOURCE: P.E. Robbins, Woods Hole Oceanographic Institution, presented by Dr. Michael Meredith.

Proxy Observations

A central point of discussion during the workshop was the need to extend the record back before the late 1970s, given the high degree of internal variability in the observational record. Examples of proxy observations that could be used to extend the observation record include whale catch locations at the ice edge (e.g., Ackley et al., 2003; Cotté and Guinet, 2007; de la Mare, 1997, 2009), ice core methanesulfonic acid (MSA) records (Abram et al., 2010; Curran et al., 2010), ship log-book records (e.g., Ackley et al., 2003; Mackintosh, 1972), ice core deuterium excess data (Sinclair et al., 2014), and Orkney Islands fast ice duration (from 1903; Murphy et al., 1995, 2014).

Such observations indicate more extensive ice prior to the early 1970s, and the recent years of increased sea ice are not unique, said many participants. Dr. Will Hobbs, University of Tasmania, noted that proxies depict an unusual phenomenon in the Ross Sea, with sea ice extent decreasing in the 1950s and 1960s followed by an ongoing increase. A major challenge of proxy reconstructed time series is the uneven geographical representation and uncertain accuracy, which is difficult to quantify and validate. Dr. Stephen Ackley, The University of Texas at San Antonio, noted that proxy data on ice extent may be difficult to compare with passive microwave extent estimates based on the 15% contour. However, several participants said proxy data are invaluable and show consistency in results and findings.

Dr. Sarah Das, Woods Hole Oceanographic Institute, presented work on developing sea ice proxies from ice cores. Ice cores are comprised of preserved annual snow layers; measuring their physical and chemical variables can provide clues about the past. For example, ice layers preserve marine-derived aerosols. Many of these aerosols are related to sea ice processes, which allows scientists to measure changes in concentration over time to estimate past sea ice.

MSA is derived from phytoplankton that thrive in the ice marginal zone and is the most promising proxy (Abram et al., 2013). One of the earliest studies compared ice-core MSA with total Southern Ocean sea ice area from the mid-1970s to the late 1980s (Welch et al., 1993; see Figure 5). The decline of sea ice area matches well with the decline of MSA, and MSA also matches regional ice extent (ice edge) in several locations (e.g., Curran et al., 2003). But, Dr. Das said, there are some areas that have an opposite/negative relationship between MSA and sea ice extent. It would be useful to combine records from different sites to get a good overall picture. In general, the proxy data suggest that there is a longer-term decline in sea ice extent and area, said Dr. Das.

Dr. Das highlighted some other new potential sea ice proxies records from ice cores:

- Halogens (Br and I) for both Arctic and Antarctic regions.
- Sulfur isotopes for deconvolving source type (e.g., polynyas versus ice edge) based on different transport behavior.
- Parameters derived from stable water isotope measurements (e.g., deuterium excess) because they are strongly influenced by sea ice cover in the moisture source region.
- Snowfall accumulation variability because it is strongly influenced by atmospheric transport, which likely also drives concurrent changes in regional sea ice.

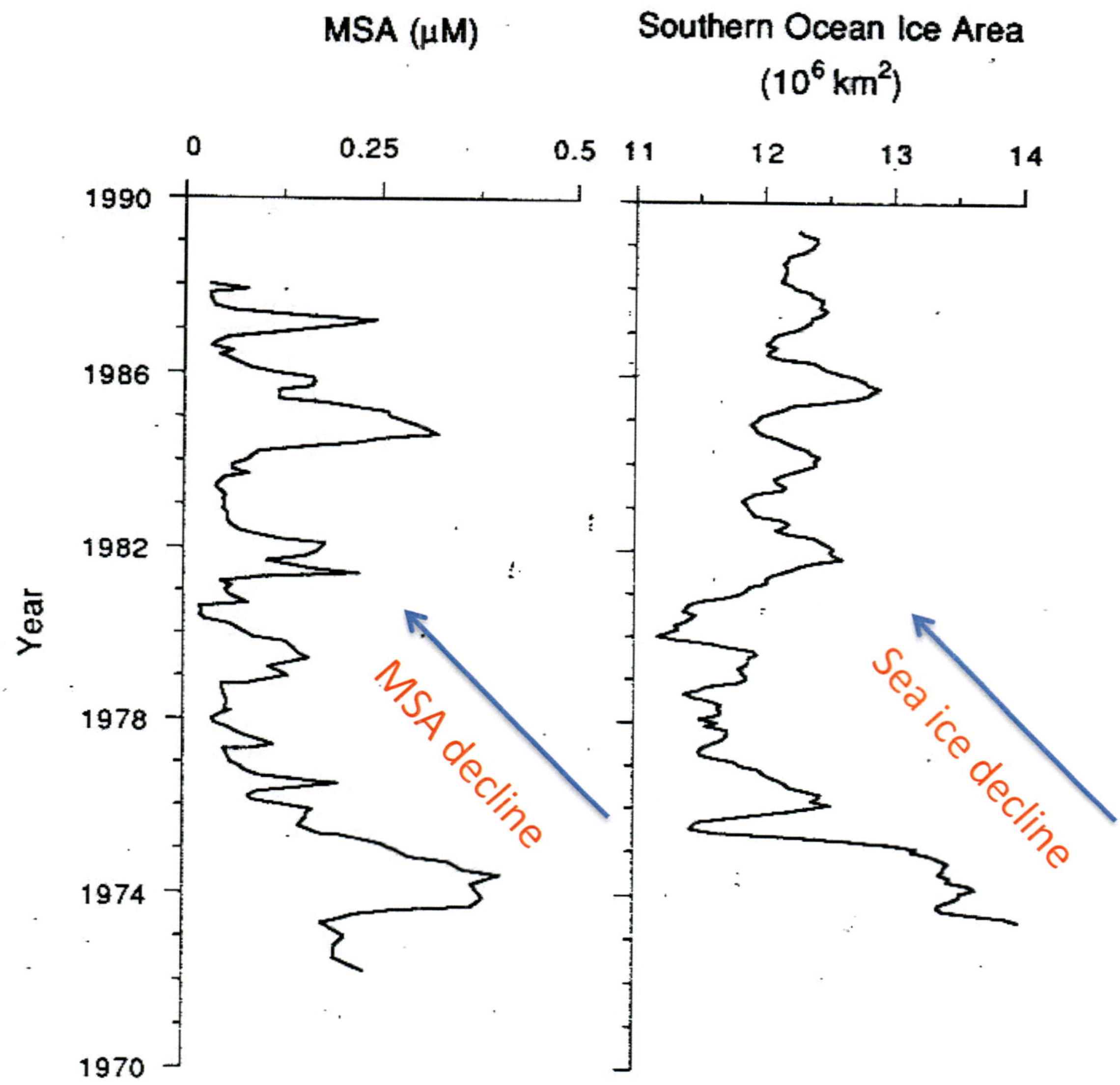

FIGURE 5 Early ice core–based reconstructions compared MSA from the Newall Glacier snowpit and total Southern Ocean sea ice area. Given the large amplitude in seasonal change in sea ice area compared to year-to-year variability, a 12-month running mean was used to smooth the sea ice record in order to visually compare it to the record of the MSA. SOURCE: Welch et al., 1993.

- Marine sediment cores because microfossils are often associated with sea ice presence or absence (e.g., diatoms, forams).
- Coraline algae, which can be found on underwater rocks throughout the world's oceans, because it is correlated with sea ice cover (presence/absence of light; Halfar et al., 2013).

Dr. Das said that one of the biggest challenges is that many factors can influence concentration of key constituents that can be measured in ice cores (e.g., MSA and sea salts). In other words, sea ice is just one of several factors. For example, the same ice core is not always reflecting moisture from the same ocean source region. Other complicating factors include aerosol production, transport, deposition, and post-depositional processes.

Reanalysis

Dr. David Bromwich, The Ohio State University, said that the Antarctic sea ice zone is almost completely devoid of conventional meteorological observations. Thus, the atmospheric parameters of most relevance to studies of Antarctic sea ice (e.g., winds, temperatures, precipitation, and downward radiation fluxes) are most often derived from global atmospheric reanalyses that merge all available observations with short-term model forecasts. Reanalysis is advancing rapidly, and the options for reanalysis datasets are growing. Most of the reanalyses start in 1979, when comprehensive satellite observations began. There are limited observation reanalyses that date back to the start of the 20th century.

Dr. Bromwich discussed trends in four sea ice–relevant variables: mean sea level pressure (MSLP), 2 m temperature, total precipitation, and sea ice conditions. For each variable, he compared several different reanalysis products and provided 1979-2009 trend maps and time series from 1900 onward.

MSLP reanalyses have a wide spread of spatial variability in trends that similarly impact the surface winds. These winds are then input into some sea ice models. In other words, the reanalysis data impact the model outcomes when models are forced by reanalyses. Based on MSLP time series at 50-65° S, all reanalyses that extend back prior to 1979 exhibit a long-term decrease in pressure, which exists regardless of the type of data assimilation.

The reanalysis products for 2 m temperature trends (1979-2009) have major differences, some of which are due to issues with station data (e.g., ERA-Interim[8]). Likewise, reanalysis products for total precipitation trends have a considerable spread in magnitude and trends.

Some reanalyses prescribe unrealistic sea ice conditions (i.e., concentration and annual mean sea ice area), said Dr. Bromwich. Sea ice conditions based on various climatologies are used prior to 1979. In some cases (e.g., ERA-20C[9]), these climatologies can depart significantly from post-1979 sea ice conditions, creating unrealistic trends in various fields.

It is challenging to use current reanalyses to analyze climate across the 20th century in high southern latitudes. The transition across 1979 is still clearly detectable in recent reanalyses, said Dr. Bromwich. He noted that the UK Met Office Hadley Centre for Climate Science and Services plans to digitalize historical records of Southern Hemisphere sea ice.

REGIONAL AND SEASONAL VARIABILITY IN ANTARCTIC SEA ICE

Observations of Antarctic sea ice show that there is significant temporal and spatial variability in the extent, which may be one reason why a simple explanation for the slight overall positive trend in the passive microwave satellite record of Antarctic sea ice extent has proven to

[8] See http://www.ecmwf.int/en/research/climate-reanalysis/era-interim.
[9] See http://www.ecmwf.int/en/research/climate-reanalysis/era-20c.

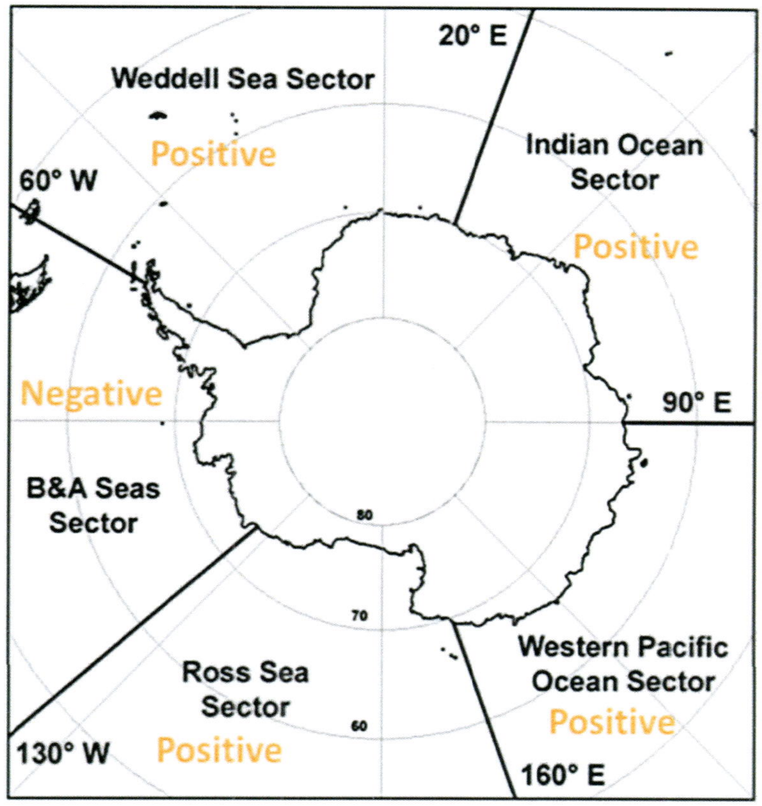

FIGURE 6 The direction of trends in monthly Antarctic sea ice extent by sector (1979-2015), based on data from the passive satellite record. SOURCE: Presented by Dr. Claire Parkinson.

be elusive. Dr. Sharon Stammerjohn of the University of Colorado Boulder said that the oft-cited weakly positive trend in circumpolar sea ice extent is actually the result of large and strongly contrasting regional trends, some with magnitudes equivalent to those observed in the Arctic. In particular, there have been strongly negative monthly extent trends in the Bellingshausen and Amundsen Seas and strongly positive trends in the Ross Sea sector (Stammerjohn et al., 2008; see Figure 6). Such regional variability is not confined to annual sea ice extent or traditionally defined sectors (Raphael and Hobbs, 2014). For example, the length of the sea ice season has changed dramatically in some areas (e.g., becoming shorter in the Bellingshausen Sea and longer in the Ross Sea; Stammerjohn et al., 2012), and there are areas of strongly contrasting trends within the traditionally defined Weddell Sea sector (Holland, 2014; Maksym et al., 2012).

Dr. Paul Holland, British Antarctic Survey, said the sea ice regional trends (i.e., area and concentration) also vary by season. The Ross Sea has a significant sea ice trend year round. No other region has a significant trend in winter and spring. In the Amundsen-Bellingshausen Seas, there is less ice in summer and no change in winter. In the Weddell Sea there is a strong increase in summer that disappears in winter (Holland, 2014).

One way to understand the seasonality of regional sea ice is to measure the seasonal rate of change of ice area (i.e., rate of ice expansion and contraction), said Dr. Holland. Expansion and contraction occur through both ice dynamics and thermodynamics. In autumn, the sea ice grows relatively slowly and in spring, the ice melts rapidly. A change in forcing (e.g., a warming) will not

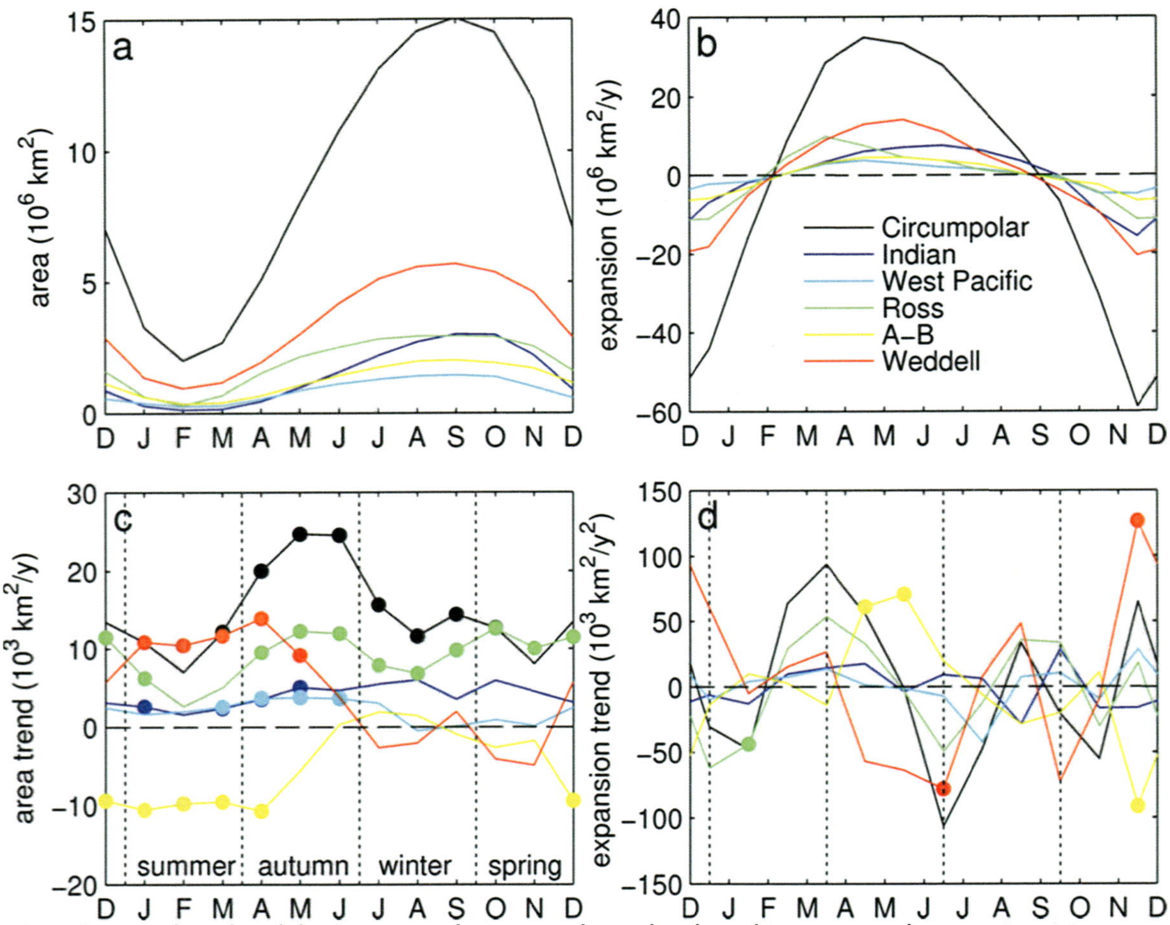

FIGURE 7 Seasonal cycle of the interannual mean and trends of total ice area and expansion: (a) mean monthly total ice area, (b) mean monthly total ice expansion, (c) interannual trends in monthly total ice area, and (d) interannual trends in monthly total ice expansion. Ice expansion in (b) causes seasonal variation in ice area in (a), and ice expansion trends in (d) cause seasonal variation of ice area trends in (c). Circles depict trends significant at 90% or above, and vertical lines illustrate seasonal boundaries. The mean was calculated using 1979-2012 quasi-daily sea ice concentration from passive microwave satellite measurements. SOURCE: Holland, 2014.

directly affect ice concentration; it will affect the rate of ice expansion and contraction, according to Dr. Holland (see Figure 7).

In looking at how ice concentration and ice expansion and have changed over time, Dr. Holland said it is clear that ice concentration trends are being created in spring and reversed during autumn. For example, there is a large increase in springtime ice loss in the Amundsen-Bellingshausen Seas, which causes a lower ice concentration to persist throughout summer until autumn, when the ice expands faster and compensates this loss. Conversely, in the Weddell Sea there is a decrease in ice loss during spring and an increase in ice gain in autumn.

According to Dr. Stammerjohn, the spatial and temporal variability points to the enormous analytical challenges associated with understanding Antarctic sea ice variability, and to the importance of finer-grained analyses to understand sea ice and its fate in the Southern Ocean.

Dr. Stammerjohn presented an analysis of sea ice extent in 2012-2014. These are years that broke records for maximum sea ice extents in the Southern Hemisphere, contrasting strongly

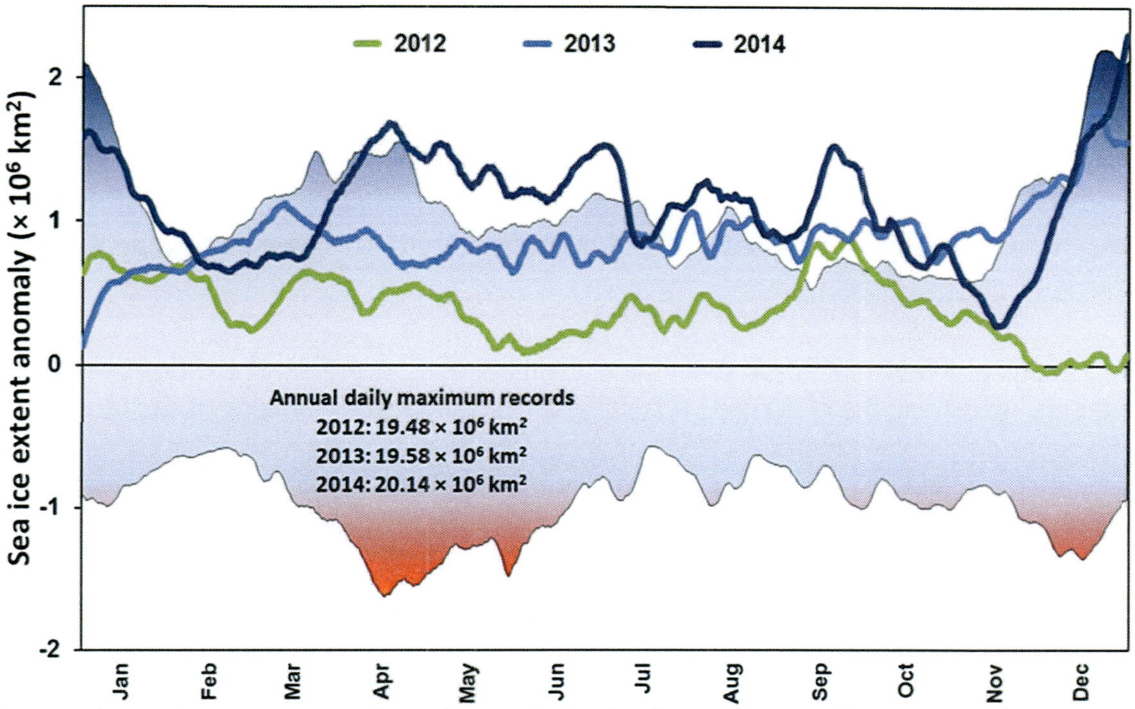

FIGURE 8 Daily sea ice extent anomaly and range in the Southern Ocean, with respect to the 1981-2010 mean, with daily departures from the mean in green for 2012, light blue for 2013, and dark blue for 2014. The shaded banding represents the range of daily values for 1981-2010. SOURCE: Presented by Dr. Sharon Stammerjohn, figure modified from Reid and Massom, 2015. ©American Meteorological Society. Used with permission.

with the record-breaking negative anomalies in the Arctic (see Figure 8). Despite the fact that these years were unprecedented in the modern satellite record, it has been very difficult to establish clear links to any one mechanism driving the successive record-breaking years because they were characterized by strong regional and seasonal differences.

Furthermore, different regions in different seasons contributed to the Antarctic-wide anomalies during the seasons of both sea ice advance and retreat,[10] and in many cases there were contrasting seasonal anomalies in the different regions. For example, in 2014, conditions in the western Weddell Sea contributed heavily to (austral) summer and early fall anomalies, while several sectors contributed to late fall and winter anomalies. Positive anomalies in the Ross Sea and Indian Ocean sectors contributed to records in winter through early spring, and large positive anomalies in the Weddell and Ross Seas led to positive anomalies late in the year (late spring and early summer).

Dr. Stammerjohn said that a very promising method to furthering understanding linkages between climate and ocean conditions and sea ice variability is to use a data-driven approach for defining regions and seasons for analyses. A recent example defined regions and seasons from the perspective of the variability in the sea ice record itself, rather than the atmosphere or the ocean, and used this as the starting place for comparison with potential drivers of sea ice

[10] Varies slightly by region, but generally considered February/March to August for advance, and October/November to February for retreat.

variability (Raphael and Hobbs, 2014). This type of analysis has yielded tighter correlations between the sea ice variability record and atmospheric phenomena during critical periods of sea ice advance and retreat, with climate modes identified as predominant proximal drivers of advance and retreat in different regions.

USING CLIMATE MODELS TO INTERPRET AND EXTEND THE OBSERVATIONAL RECORDS

Climate models provide an important tool for interpreting and extending understanding of Antarctic sea ice observations, said many participants, as well as exploring the potential mechanisms influencing the observed variability. Box 3 provides descriptions of the models that have been used to study Antarctic sea ice and the sorts of experiments that can be performed. According to several participants, much of the research to date has focused on understanding the notable discrepancy between observations and models, which generally exhibit a decline in Antarctic sea ice extent over the last 30-50 years. If the models are improved to the point that they can reliably reproduce past sea ice conditions, then they could also be used to disentangle the roles of internal variability and human-caused drivers of changes in sea ice, as well as to project how sea ice might change in the future.

Reconciling Models and Observations

The modest positive trend in satellite-derived estimates of sea ice extent observed from 1979-2014 does not match global climate model simulations of Antarctic sea ice (see Figure 9). Most models show Antarctic sea ice extent declining on average in response to observed warming of the atmosphere (Gagné et al., 2015; Mahlstein et al., 2013; Polvani and Smith, 2013; Swart and Fyfe, 2012; Turner et al., 2013b; Zunz et al., 2013).

Furthermore, said Dr. Hobbs, very few Coupled Model Intercomparison Project Phase 5 (CMIP5) models are able to replicate the observed summer sea ice extent mean state, and they tend to overestimate its year-to-year variability (Zunz et al., 2013). Dr. Hugues Goosse, Université Catholique de Louvain, said it is critical that models represent the mean state of the system accurately because several proposed mechanisms for sea ice variability depend on the mean state (Lecomte et al., 2016). Furthermore, models also need a correct initialization of the Southern Ocean to ensure more accurate decadal predictions and projection.

Over the satellite era, the models (including the CMIP archive and the ensemble mean) do not capture the cooling surface temperature trend that is occurring over the Southern Ocean (Marshall et al., 2014). However, when observations are extended back to 1940, they indicate that there has been warming in the Antarctic (though less than in the Arctic), which is consistent with the models. According to Dr. Cecilia Bitz, University of Washington, scientists should thus reframe the question away from expecting the ensemble mean of models to resemble the observed estimate of Antarctic sea ice extent. The goal could be for modeled response to GHG to match the longer-term record rather than the satellite record. Perhaps the satellite record represents internal variability or shorter-term forcing, not well represented in models.

BOX 3
Introduction on Modeling Antarctic Sea Ice Variability

Dr. Clara Deser, National Center for Atmospheric Research (NCAR), provided an introduction on the many different types of models and experiments that can be used to study the processes impacting Antarctic sea ice extent. One type is free-running coupled climate models. Such models include components for the atmosphere, ocean, sea ice, land, and biology. They are scientists' best attempt to replicate the climate system. The CMIP5 models used in IPCC Assessment Reports and NCAR's Community Earth System Model (CESM) are examples of coupled climate models.

Control simulations using coupled climate models show the internal variability of the model's climate system (e.g., ENSO, decadal variability) in the absence of changes in radiative forcing. They are long simulations of at least 1,000 years for robust sampling. Historical simulations simulate the past, typically the 20th century, and have internal variability. They respond to changing radiative forcing and any prescribed forcing (solar irradiance, GHG, aerosols, and stratospheric ozone; Deser et al., 2012). Historical simulations tend to be used for model evaluation, which is the process of evaluating climate sensitivity, internal variability, and the response to external forcing. Historical runs also provide information on the range of possible outcomes. In comparing the historical runs to observations, the observations should fit within the range of simulations. However, no single simulation needs to represent the actual observed climate (even if the model were perfect), because simulations have different chronologies of decadal and multidecadal variability.

Constrained coupled climate models are constrained to include the observed variability in a certain area and how that impacts other areas. Simulations from these constrained models are referred to as "Pacemaker" simulations because the observed variability in a particular area sets the "pace" for the rest of the climate system. These models are particularly good for hypothesis testing and physical and mechanistic understanding. For example, to test the influence of the tropics on the Southern Ocean, a scientist could "nudge" winds or SSTs in the tropics toward observations and allow the free-running coupled model to respond (Douville et al., 2015; Kosaka and Xie, 2013; Schneider et al., 2012a).

Antarctic sea ice can also be studied with "forced-component models." These sorts of model experiments allow scientists to test hypotheses and gain process-level understanding. For example, to explore the influence of observed wind changes on the Southern Ocean, a scientist would prescribe the observed evolution of surface winds to a coupled ocean–sea ice model. However, these models do not provide information on the origin of the wind changes (Bintaja et al., 2013; Sen Gupta and England, 2006; Zhang, 2014).

Dr. Xiaojun Yuan, Lamont-Doherty Earth Observatory, discussed a statistical forecast model that is capable of capturing the Antarctic sea ice response to ENSO forcing. This model depicts large interannual variability in sea ice concentrations in the Western Hemisphere. Sea ice in the region is quite responsive to atmospheric forcing at a seasonal time scale, demonstrating that sea ice anomalies are predictable with a simple statistical model. Dr. Elizabeth Hunke, Los Alamos National Laboratory, said that the MPAS (Model for Prediction Across Scales)–sea ice,[a] which is a component of a new climate model from the Department of Energy, is being used to study ice-ocean-atmosphere interactions in the Southern Ocean.

[a] See https://mpas-dev.github.io.

Dr. Bitz noted that Fan et al. (2014) found a strong relationship between SST and sea ice extent, surface wind, and sea level pressure when observations are extended back to 1940. Armour and Bitz (2015) show that sea ice extent, SST, and westerly wind anomalies are all correlated (see Figure 10).This indicates that there has potentially been a loss of sea ice in

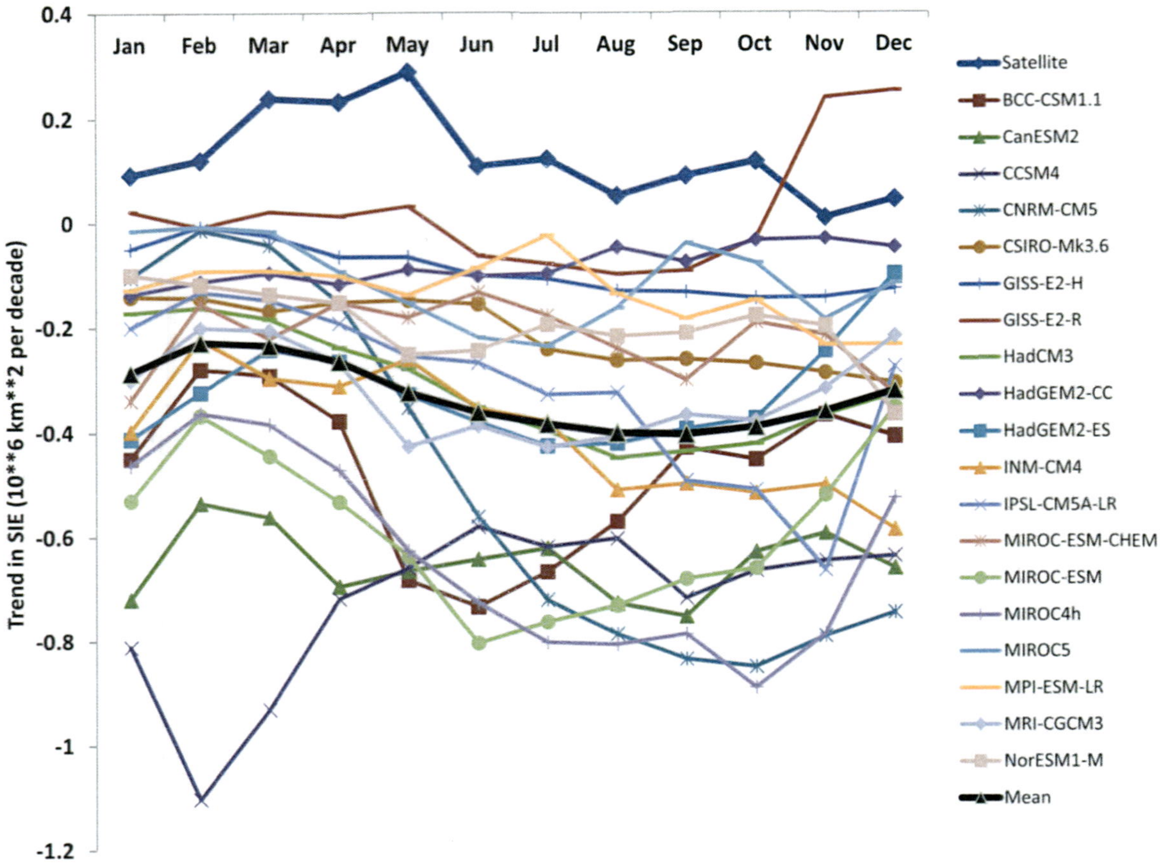

FIGURE 9 Monthly trends of sea ice extent from the satellite data and CMIP5 models over 1979-2005 shown in absolute trend per decade (106 km^2 $decade^{-1}$). The mean of the ensemble members is plotted for models with multiple ensemble members. The black line depicts the mean of all the models; the thick blue line depicts the mean of satellite observations. SOURCE: Turner et al., 2013b. © American Meteorological Society. Used with permission.

Antarctica over the century scale, along with a slight increase in SSTs, which is also shown by CMIP models on the century scale. Thus, there is a better match between models and observations on longer time scales, in terms of both surface air temperature, SST, and quite possibly sea ice extent.

Dr. Lorenzo Polvani, Columbia University, said it is important to recognize that the models simulate the sea ice extent seasonal cycle quite well (Zunz et al., 2013). Furthermore, many models even capture the regional features correctly. However, some participants said that models show larger-scale variability and cannot provide information about regional variability.

Dr. Laura Landrum, NCAR, said that observations of the trends in sea ice extent over the seasonal cycle are well within the uncertainty of CMIP class models over much of the Southern Ocean. However, there are several exceptions. For example, the Ross Sea (especially the western Ross Sea) during January through May is not well captured in the models. Several reasons for this discrepancy include

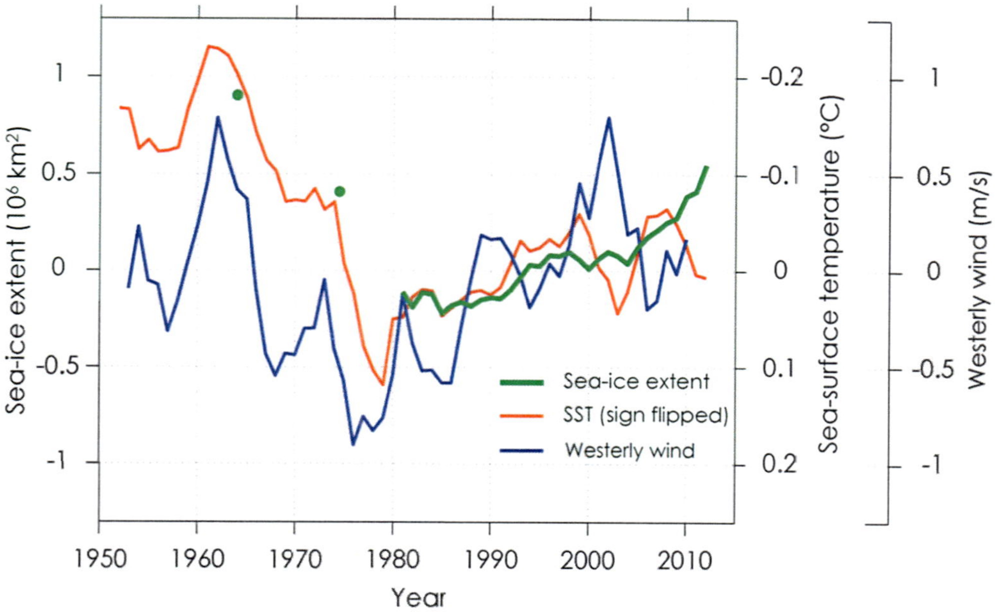

FIGURE 10 Time series of anomalies in the total annual mean Antarctic sea ice extent (green), annual mean Southern Ocean SST (red), and summer westerly wind (blue). All anomalies are taken with respect to their 1980-2010 means. All data are annual except 1964 sea ice is September only. SST is NOAA Extended Reconstuction; westerly wind is station data (from Fan et al., 2014); 1964 sea ice is from Meier et al., 2013; and the sea ice extent in 1974 is an average of 1973-1976 from ESMR.[11] SOURCE: Armour and Bitz, 2015.

- Ice shelf processes and freshwater input;
- Polynyas, ice shelf, and continental slope generally not resolved in coupled models; and
- CESM ocean is too stratified on the continental shelf of the Ross Sea.

The Weddell Sea in summer is also not well captured in the models, but to a lesser degree than the Ross Sea.

Dr. F. Alexander Haumann, ETH Zürich, said that the differences in observed and simulated Antarctic sea ice variability may be caused by different atmospheric circulation patterns and ocean stability (Haumann, 2011; Haumann et al., 2014). Zonal asymmetries in the large-scale circulation around Antarctica drive variations and long-term trends in Antarctic sea ice, and this process is not captured well in climate models. Dr. Haumann said that the underestimated zonal asymmetries in the atmospheric circulation are also causing a much weaker ocean surface stratification in the model. This results in a much higher sensitivity of the sea ice to changes in the subsurface heat fluxes, explaining why simulated sea ice changes could be driven by a different mechanism than the changes observed in recent decades.

Another reason why models are not capturing observations is due to a combination of model biases, said Dr. Armour. Such biases include

[11] See https://nsidc.org/data.

- Shortwave radiation, which strongly affects the sea ice;
- Vertical and horizontal ocean temperature gradient biases over the whole Southern Ocean, perhaps due to inability to resolve coastal processes (e.g., Antarctic bottom water formation); and
- Biases in wind trends, both Southern Annular Mode (SAM) and local circulations (up to uncertainties in observations). This could be due to a combination of incorrect ozone forcing in the models, and internal variability.

Dr. Jennifer Kay, NCAR, noted that the excessive absorbed shortwave bias over the Southern Ocean in CESM has been "fixed" by making the clouds brighter. Addressing this long-standing radiation bias results in important climate impacts for the Southern Ocean, illustrating that much can be learned about how the system works by "fixing" these large existing biases.

Many workshop participants said explaining discrepancies between models and observations would lead to improvements in the models and for understanding the future of sea ice. Although models are useful for exploring mechanisms, several participants said that models are not ready to be used for attribution.

Separating Internal Variability from Anthropogenic Forcing

During the discussion, many participants noted that the discussion of the drivers of Antarctic sea ice extent variability has been framed as either anthropogenic forcing or internal variability. However, the expansion of sea ice could be a combination of both, so dominant mechanisms should be emphasized rather than forcing or internal variability, according to several participants.

For detection and attribution, scientists should move away from the idea that the observed climate is the "right" one, said Dr. Hobbs. The observed climate is one realization of the climate. Studies should frame the climate as signal and noise, one of an ensemble of possible climates that could be realized under a given forcing. The difficulty with studying mechanisms and drivers of Antarctic sea ice is that the signals are all in areas of high noise. From a detection and attribution perspective, it makes more sense to study annual mean sea ice extent to get a signal-to-noise ratio. However, important information is lost due to averaging.

If 1979-2005 trends from 100 ensemble members of CMIP5 model runs are averaged, it is difficult to say anything definitive about the models (i.e., whether they are "right" or "wrong" [Maksym et al., 2012]). In CMIP5, almost all of the sea ice response is driven by GHGs; there is some concern that the ozone depletion in the models is not representative, said Dr. Polvani. He emphasized that disagreements between observations and multimodel mean does not imply that models are fundamentally flawed. Given that the multimodel mean comes from combining many different model simulations, the natural variability is "averaged out" and therefore represents the forced response only (Polvani and Smith, 2013). Thus, the multimodel mean and observations will agree only if internal variability is a relatively small driver of the Antarctic sea ice variability, according to Dr. Polvani.

Dr. Hobbs and Dr. Polvani said this suggests that internal variability is large compared to anthropogenic forcings. Many recent modeling studies have indicated that the anthropogenic signal, whether caused by increasing CO_2 or decreasing stratospheric ozone, would result in a reduction in ice cover and may be relatively small compared to internal variability. No one mode or process can explain the entirety of the trends. The atmosphere and ocean both play a part, with great spatial dependence on which processes dominate. Polvani and Smith (2013) showed that both forced and observed trends fall well within the probability density function of the preindustrial control runs. Gagné et al. (2015) also found that the recent increase in Antarctic sea ice is due to internal variability.

Dr. Polvani also noted that observations of Antarctic sea ice extent, SSTs, and zonal winds appear to be largely uncorrelated with the increase in carbon dioxide and the decrease in ozone. Observations show a clear reversal of trends around 1980, whereas both major anthropogenic forcings are monotonic since 1950; hence, trends are unlikely to be primarily due to anthropogenic forcings.

Projecting Future Antarctic Sea Ice Conditions

Some participants said that Antarctic sea ice is likely to decline in the longer term, which is in line with model projections. However, others said that models are not ready to make confident projections about 21st-century Antarctic sea ice and that scientists need a better understanding of the mechanisms before confident statements can be made about the future of Antarctic sea ice. For example, if freshwater (e.g., due to ice sheet melt and enhanced precipitation) is really a significant driver of sea ice expansion, perhaps that will continue to dominate GHG trends in the future. On the other hand, if winds have driven expansion, future wind changes (forced or internal variability) may dominate.

Many participants said that in the long term (~2,100) models likely get the right sign of the response to the forcing (i.e., Antarctic sea ice will likely decrease). The Southern Ocean will probably get warmer with thermodynamic consequences for the sea ice, but it is difficult to provide accurate information about locations and magnitude of sea ice trends.

Many participants highlighted the importance of disaggregating overall trends to extrapolate regional variability—with distinct controlling mechanisms—as the drivers of overall trends. Dr. Povlani reminded participants that looking regionally increases signal to noise, whereas averaging hemispherically creates small trends and decreases signal to noise. The largest warming and sea ice loss occurs in the Bellingshausen-Amundsen Seas; this is the only region where observations are in line with what the models predict for the forced response. Yet, this is where some scientists think internal variability has played a primary role in the changes.

PROPOSED MECHANISMS DRIVING OBSERVED VARIABILITY IN ANTARCTIC SEA ICE

There are many local, regional, and global processes that influence seasonal sea ice growth and melt, including the internal variability of the coupled atmosphere–ocean–sea ice system as well as external forcing to processes (e.g., indirect effects of stratospheric ozone depletion, and greenhouse gases). Workshop presentations and discussions highlighted several individual processes and mechanisms that could influence recent trends and record extents in Antarctic sea ice (see Box 4). These processes and mechanisms are not necessarily mutually exclusive. Many participants said that because every region of the Antarctic is sensitive to different climate anomaly patterns, no one mode or mechanism can be evoked to explain the recent overall record extents.

Feedbacks and Linkages Between Atmospheric Warming and Local Hydrology, Sea Ice Formation, and Ocean Circulation

Sea Ice–Ocean Feedbacks and Linkages

Sea ice feedbacks and linkages can result in opposite drivers leading to the same outcome, said Dr. Hobbs. For example, increased sea ice cover (Goosse and Zunz, 2014) leads to increased spring melt and increased freshwater input, which results in a strongly stratified mixed layer and reduced ocean flux, favoring sea ice formation. Atmospheric warming, on the other hand, leads to reduced autumn ice production and a reduced brine rejection that also results in a strongly stratified mixed layer and reduced heat flux, which also favors sea ice formation (Zhang, 2007).

Dr. Doug Martinson, Lamont-Doherty Earth Observatory, emphasized that the ocean–sea ice negative feedback keeps Antarctic sea ice thin (and thus seasonal). The winter evolution of the water column is the most fundamental process in the Southern Ocean that works to control the stratification; the nature and timing of the sea ice distribution; and deep and bottom water formation, productivity, and ventilation of heat and CO_2. The ice-ocean feedback describes how many units of ice would melt for every one unit of ice grown due to entrainment of heat from brine rejection (Martinson and Iannuzzi, 1998). For the Atlantic sector (Weddell gyre) of the Antarctic sea ice fields, each unit grown in one day results in enough heat released to melt 4 to 16 units of ice, or alternatively warms the water such that it takes 4 to 16 units (i.e., 4 to 16 days) to vent the heat before more ice can grow. Hence, very little ice can grow over several months of winter.

Waters capable of growing sea ice are limited by the ACC, so slight changes in its position due to variability in westerlies can set the average northern limit of sea ice fields. The ACC divides polar from tropical waters and limits the equatorward expansion of ice growth, said Dr. Martinson. Near the ACC, the negative feedback gets stronger and stronger, resulting in thinner ice formation that cannot survive the melting from the warm surface waters while being blown northward, thus limiting that form of expansion as well. It is speculated that subtle shifts in the ACC via changes in the westerlies may actually account for the statistical zonal sea ice expansion. Dr. Lynne Talley, Scripps Institution of Oceanography, noted that topography also constrains the southern ACC fronts.

The Southern Ocean has also freshened, which is consistent with an accelerating hydrological cycle (Böning et al., 2008). Dr. Haumann discussed research on estimating uncertainties and freshwater fluxes associated with sea ice melting, sea ice freezing, and sea ice transport. This research depicted a persistent and robust trend of freshwater advection due to

BOX 4
Possible Mechanisms of Antarctic Sea Ice Variability[a]

Feedbacks and linkages between atmospheric warming, hydrology, sea ice formation, and ocean circulation:

- Atmospheric warming → accelerated melting of Antarctic ice shelves → increased freshwater input → more strongly stratified mixed layer → reduced ocean flux from deep ocean to surface waters → cooler, fresher sea surface conditions → enhanced sea ice formation (Bintanja et al., 2013).
- Increased sea ice cover → increased spring melt of sea ice → increased freshwater input → more strongly stratified mixed layer → reduced ocean flux from deep ocean to surface waters → cooler, fresher sea surface conditions → enhanced sea ice formation (Goosse and Zunz, 2014).
- Increases in precipitation → freshening of surface waters → cooler, fresher sea surface conditions → enhanced sea ice formation (Liu and Curry, 2010; Zhang, 2007).

Variability in regional and large-scale atmospheric circulation and surface wind patterns, including

- The Southern Annular Mode (SAM) and in the strength and position of associated westerly (zonal) winds (Lefebvre et al., 2004; Simpkins et al., 2012; Stammerjohn et al., 2008).
- The Amundsen Sea Low (ASL; Fogt and Scambos, 2013, 2014; Fogt and Zbacnik, 2014; Matear et al., 2015; Turner et al., 2015, 2016).
- Zonal wave 3, one of the quasi-stationary waves within the strongly zonal high-altitude atmospheric winds in the Southern Hemisphere high latitudes (Raphael, 2007; Raphael and Hobbs, 2014).

Influence of stratospheric ozone depletion (or the "ozone hole") on sea ice variability through its impact on

- Westerly winds and the Southern Annular Mode (Thompson and Solomon, 2002; Thompson et al., 2011; Turner et al., 2009, 2015);
- Feedbacks between winds, ocean circulation, and sea surface temperature (Armour and Bitz, 2016; Ferreira et al., 2015; Goosse et al., 2009; Kostov et al., 2016; Marshall et al., 2014); and
- Strength and position of persistent low-pressure systems (Haumann et al., 2014).

Connections between tropical ocean variability and sea ice:

- Links between Antarctic sea ice and the El Niño–Southern Oscillation (ENSO) and the Interdecadal Pacific Oscillation (Ding et al., 2011; Meehl et al., 2016; Schneider et al., 2012a; Yuan, 2004; Yuan and Martinson, 2000).
- Connections to the Atlantic Multidecadal Oscillation (Li et al., 2014; Simpkins et al., 2014).

Regional and/or seasonal processes affecting storminess and winds, which influence air and ocean temperature, sea ice production and export, and increases in sea ice thickness and extent in certain areas (Holland and Kwok, 2012; Holland et al., 2014).

[a] This is not a comprehensive list of all possible mechanisms of Antarctic sea ice variability. Elements of these mechanisms were discussed at the workshop and summarized in this proceedings.

changes in the drift of sea ice from the continent to open water. According to Dr. Haumann, sea ice freshwater flux changes may explain effects on salinity changes in the Southern Ocean in the recent decade. A model sensitivity study was conducted to examine the impacts on salinity, with the addition of glacial melt along the coast of the Amundsen Sea leading to overcompensation and subsequent freshening linked to a feedback in sea ice transportation trends. The resulting conclusions determined that salinity changes and freshening in the Southern Ocean are driven by changes in sea ice transport.

An important question, said Dr. Meredith, is whether the ocean overturning has accelerated. Analysis of tracer data indicates that the Southern Ocean overturning may have accelerated (Waugh et al., 2013), and this potentially represents an increase in upwelling of warm water from below. However, this is very difficult to measure, and current understanding is based on very few data points. Furthermore, for any upwelled warm water to influence sea ice, it must cross the pycnocline and penetrate the upper mixed layer; this may be difficult if stratification is stronger. According to Dr. Meredith, it is very difficult to attribute strengthened Southern Ocean stratification to either ocean-induced melting of ice shelves or increases in precipitation. Increases in glacial discharge in some sectors are fairly robust, and circumpolar spreading around the Antarctic shelf has been shown using satellite data; however, it is not clear whether this freshening on the surface has significantly impacted sea ice production (Rye et al., 2014).

Ice Sheet–Ocean–Sea Ice Feedbacks

Warming of global ocean deep water has considerable potential for long-term melting of Antarctic glacial ice shelves. Although the Southern Ocean is not well measured in all places, there is evidence that the Upper Circumpolar Deep Water (UCDW) has been warming dramatically, according to Dr. Martinson. The UCDW is the warm deeper water responsible for melting the ice shelves. Rhein et al. (2013) found that the deep Southern Ocean is warming significantly faster than the global ocean as a whole. The warming ensures strong melt, even if the circulation beneath the ice margins slows down.

To better understand impact of ice sheet melt on sea ice, several model studies have been conducted. Dr. Ryan Walker, NASA, discussed a 2004 modeling study run by Hellmer (2004) that used ice-ocean models run with and without the inclusion of major subshelves in the ocean. It showed that meltwater contributes to sea ice increases in two ways: through (1) cooling and freshening of surface waters and (2) stabilization of the water column. This demonstrates that the spatial pattern of sea ice response depends on dynamics of ocean currents.

Bintanja et al. (2013) conducted a modeling study where they simulated processes associated with increased ice sheet melt in a coupled climate model to explore how those processes impacted sea ice extent. They used best estimates at the time of grounded ice sheet mass imbalance (250 Gt/yr loss for 31 years; latest estimates are one-third as much), and the sea ice expanded. They argued that, zonally around the continent, there is on average cooling and freshening in the surface layers, warming in the deeper layers, and an increase in stability. This phenomenon is due to sub–ice shelf melting and is considered by Bintanja et al. to be the dominant driver in increased sea ice, more so than any atmospheric factors. However, Swart and

Fyfe (2013) and Pauling et al. (2016) found that the sea ice response was small when they added a similar amount of freshwater to their models.

Pauling et al. (2016) found that the rate of rain that is falling in the Southern Ocean is about 25,000 Gt/yr. This is more than 200 times greater than the rate of freshwater Bintanja found was sufficient to cause sea ice to expand. Pauling et al. added 2,000 Gt/year of freshwater to the CESM model. Sea ice extent increased but not enough to change the trend, said Dr. Bitz.

Different answers from different models reflect model biases, likely biases in the mean state, and especially errors in their climate sensitivity, according to Dr. Bitz.

Variability in Regional and Large-Scale Atmospheric Circulation and Surface Wind Patterns

Southern Annular Mode

The SAM is the dominant mode of atmospheric variability in the Southern Hemisphere and is thought to influence sea ice via its influence on westerly winds. Circumpolar winds become stronger during the positive phase of the SAM as pressure gradients between the middle and high southern latitudes increase, driving increases in the strength of westerly winds. This can cause expansion of sea ice in some regions due to enhanced Ekman drift in the ocean that favors northward advection of sea ice (Stammerjohn et al., 2008). Dr. Hobbs noted that increased westerlies can also lead to a decrease in sea ice because the Ekman transport brings up warm, deep water that can melt the sea ice. The SAM has been experiencing a positive trend since the mid-1960s and is forced by GHGs and ozone.

Dr. John Fyfe, Environment and Climate Change Canada, said that model studies show that Southern Hemisphere surface westerlies account for about 25% of the Antarctic sea ice anomaly. Models participating in CMIP5 suggest that externally forced strengthening of the westerlies may have impacted Antarctic sea ice coverage. There is little agreement between models and reanalyses (1950 to the present) in terms of the strength and trends in the westerlies, but there is a zonal pattern.

However, from 1979 to the present, the strength of the trend in the westerlies for the austral summer is fairly consistent. All of the other seasons span from negative to positive trends. The annual mean does not exhibit a shift because the winter shift is balanced by the summer shift. Patterns of change from 1988 to 2011 show no zonal structure; rather, it is a dipole pattern, suggesting a larger role of internal variability over that time period. Dr. Fyfe noted that the Antarctic trend in the last 7 years is outside of anything observed and the pattern of change is dissimilar from the CMIP5 average pattern. He concluded that an externally forced westerly wind impact on Antarctic sea ice coverage is plausible, but solid evidence has yet to emerge. Further progress has been hindered by the lack of long-term observations, the wide range of reanalysis estimates, the influence of internal variability, and the systematic bias across models.

Amundsen Sea Low

Another atmospheric driver of Antarctic sea ice variability is the ASL. The ASL controls meridional winds near West Antarctica (in the Amundsen, Bellingshausen, and Ross Seas), as well as sea ice motion. Dr. Hobbs said that spring trends in wind and sea ice concentration from 1992 to 2010 show that a deepening of the ASL is bringing warm air into the Bellingshausen Sea and cooler air over the Ross Sea. However, when the data are extended back to 1979, this trend is not apparent. The location of the ASL has shifted and also does not line up with the increases in sea ice concentration in the western Ross Sea. Thus, said Dr. Hobbs, he is skeptical about a connection between the increase in sea ice concentration over the western Ross Sea and the ASL.

Dr. Fogt noted that the ASL deepened in every season except winter over 1979-2008 (Turner et al., 2013a). Model studies suggest that the summer strengthening of the ASL can be at least partially attributed to ozone depletion (Turner et al., 2009; the role of ozone is discussed in more detail below). However, the deepening in autumn appears to be model-dependent because not all model studies find this relationship between the ASL and ozone (Fogt and Zbacnik, 2014; Turner et al., 2009). Other changes in the ASL are more strongly tied to internal variability, especially tropical forcing (Fogt and Wovrosh, 2015; Raphael et al., 2016; Turner et al., 2015). The features of the ASL need to be adequately simulated in order to accurately depict regional sea ice trends.

Influence of Stratospheric Ozone Depletion

There was additional workshop discussion on the influence of stratospheric ozone on Antarctic sea ice variability through its connection to local winds. Since the 1970s, westerly winds have strengthened and moved poleward, exhibiting a positive SAM-like pattern. It is thought that the primary driver is ozone depletion (Polvani et al., 2011), but it may also reflect internal variability (Deser et al., 2012). As discussed above, increased westerlies may enhance sea ice formation or decrease it.

There is an observed correlation between ozone depletion and sea ice expansion (e.g., Goose et al., 2009). However, Dr. Polvani said that all available CMIP-class model studies specifically designed to isolate the effects of increasing chlorofluorocarbons and stratospheric ozone depletion show decreased sea ice extent over Antarctica due to warming of the surface ocean (Bitz and Polvani, 2012; Sigmond and Fyfe, 2010).

To reconcile these two points of view, Dr. David Ferreira, University of Reading, did a modeling experiment and found that ocean–sea ice response to abrupt ozone depletion has two phases in the high latitudes: cooling and then warming (Ferreira et al., 2015). Initially, the positive SAM-like pattern (forced by ozone depletion) enhances northward Ekman drift in high latitudes but weakens it in the lower latitudes. This leads to SST cooling in high latitudes and warming in lower latitudes. This phase typically lasts for a few years. Over longer time scales, this dipole response slowly evolves into a hemispheric warming; the high-latitude enhanced northward Ekman drift leads to upwelling of warm waters from below the mixed layer. This warming eventually overtakes the cooling from anomalous Ekman drift due to sustained bursts of westerly winds as a result of ozone hole depletion. The two time scales emerge through ocean dynamics. This results in sea ice expansion followed by a decrease, in phase with the high-latitude SST cooling and warming.

The time scale of the transition between the cold and warm SST responses appears very uncertain (between 3 and 20 years), with several sources of uncertainty, including air-sea fluxes (clouds/shortwave), strength and shape of the wind stress response, and ocean eddy processes (control upwelling rate). In the face of these large uncertainties, Dr. Ferreira believes that scientists should refrain from making statements about ozone depletion contributing or not to recent Antarctic sea ice trends.

The implications of the two time scales goes beyond the SAM/ozone context: The complex ocean dynamics observed in the ozone case is likely to be triggered by other sources of SAM variability (internal variability, tropical teleconnection, and CO_2 forcing) and by other modes of atmospheric variability. The two–time scale behavior also implies that the relationship between SAM and SST/sea ice is not the same on all time scales, an assumption often made when analyzing observations. For example, the ocean–ice response to a SAM trend is not necessarily a trend.

Connections Between Tropical Ocean Variability and Antarctic Sea Ice

Tropical Pacific Teleconnections

Dr. David Schneider, NCAR, said that changes in tropical SSTs have played the dominant role in deepening the ASL during the satellite era. He believes that the tropics may be setting the pace of the changes in the Antarctic, i.e., the tropics are the "pacemaker" of Antarctic anomalies.

Dr. Schneider described a study he did using coupled simulations of Antarctic sea ice and related variables to look at this pacemaker hypothesis. He used a large ensemble of 30 members with forcings similar to CMIP5 historical experiments (up to 2005) and Representative Concentration Pathway 8.5 (after 2005). The Pacemaker ensemble is comprised of 10 members where SST anomalies are restored to observed values in the eastern tropical Pacific. He hypothesized that the inclusion of tropical SST variability in the Pacemaker runs would improve the wind trends, which would in turn improve the sea ice trends. In comparing the runs from the large ensemble and the Pacemaker runs, the Pacemaker runs simulate a richer pattern in zonal winds that captures the trade winds and is more similar to reanalysis. However, the model still loses sea ice. Dr. Schneider believes this is due to subsurface ocean processes because the atmosphere is constrained by observations.

Dr. Graham Simpkins, University of California, said that the influence of Pacific SST variability (ENSO) on Southern Hemisphere high-latitude climate has long been known (Karoly, 1989; Mo, 2000; Mo and Higgins, 1998), and there is a distinct regionality and seasonality on resulting teleconnections.

To highlight this regionality and seasonality, Dr. Simpkins discussed SST anomalies in the east Pacific (Canonical El Niño; EP) and the central Pacific (El Niño Modoki; CP) and the resulting teleconnections to the Antarctic (Ciasto et al., 2015). On interannual time scales, both EP and CP SST variability drive pronounced perturbations to the Antarctic surface climate, but with marked seasonal dependence. In the winter, EP and CP SST variability promotes very similar teleconnection structures with a westward shift in CP compared to EP. These teleconnections are characterized by a stationary Rossby wave train. Anticyclonic circulation anomalies (and associated winds) over the ASL drive anomalous sea ice variability in the West Antarctic region (Ciasto and Thompson, 2008; Ding and Steig, 2013; Karoly, 1989; Liu et al., 2004; Schneider et al., 2012b; Simpkins et al., 2012; Stammerjohn et al., 2008; Turner, 2004; Yuan, 2004).

However, in the summer, differences are observed in the atmospheric structures of EP and CP. The EP teleconnection projects onto the negative phase of the SAM, whereas the CP exhibits a weak teleconnection. The Antarctic impacts of the EP and CP are somewhat similar in terms of spatial structure, but CP anomalies are weaker. Anticyclonic circulation (and associated winds) promote a dipole pattern in sea ice concentration between the Ross/Amundsen and Bellingshausen Seas (Ding et al., 2012; Fogt and Bromwich, 2006; Fogt et al., 2011; Karoly, 1989; L'Heureux and Thompson, 2006; Simpkins et al., 2012; Stammerjohn et al., 2008).

Given these interannual relationships, Dr. Simpkins said claims have also been made regarding longer-term impacts on the Antarctic climate, driven by Pacific SST trends. These impacts are largely linked to the CP/South Pacific Convergence Zone region. A somewhat uncertain connection has been made with CP in the wintertime, but the reanalyses might not support it. In the springtime, there appear to be Pacific South American links. While Pacific teleconnections potentially exist on these longer-term time scales, they explain only a component of the trends in specific regions. There are many confounding processes, including the Atlantic.

Tropical Atlantic Teleconnections

Tropical Atlantic SSTs are another potential mechanism for Antarctic sea ice variability. According to Dr. Xichen Li, University of California, San Diego, this mechanism can be explored through comparisons of statistical analyses and coupled-model simulations. North Atlantic SSTs have increased by more than a half degree, and this is impacting the Antarctic, said Dr. Li.

Atlantic SST variability can impact the Antarctic atmospheric circulation directly through stationary Rossby waves. The Atlantic can also first influence the Pacific, which further generates the Rossby wave chains to the Antarctic. These Rossby wave trains in western Antarctica modify the regional atmospheric circulation and thus the Antarctic sea ice distribution.

Several numerical models confirm that Atlantic SST variability deepens the ASL, which impacts the Antarctic sea ice. This is a strong teleconnection, particularly in wintertime. Atlantic SSTs may also interact with SST changes in the Indian Ocean and Pacific through the Rossby and Kelvin wave dynamics.

Although these mechanisms are robust, Dr. Li noted that there are also multiple pathways for the different ocean basins to interact, which is a complicating factor in the tropical-polar teleconnection.

Mechanisms in the Context of Regional Variability

Dr. Stammerjohn said that global and continental-scale factors likely altered the probability for years with high sea ice extents (2012-2014). For example, the SAM was mostly positive, and ENSO was neutral during the 2012-2014 record years (Fogt and Scambos, 2013, 2014; Fogt and Wovrosh, 2015; Reid and Massom, 2015). In contrast, fewer regional- and continental-scale positive sea ice extent anomalies characterized 2015, a year with a strongly positive El Niño. Despite the fact that these processes are known to influence sea ice variability, Stammerjohn emphasized that the state of SAM and ENSO cannot explain variability in all seasons and regions and thus Antarctic sea ice variability more generally (Liu et al., 2004; Yuan and Li, 2008). For example, they cannot explain all of the variability contributing to the record high sea ice anomalies in 2012-2014. Instead, the influence of local and regional climate patterns and their influence on local wind and ocean conditions were critical:

- Positive sea ice anomalies in the Weddell Sea were associated with preconditioning by cool SST anomalies that persisted from the previous year, driving very early advance of the sea ice. These anomalies in the Weddell Sea may have also influenced later positive anomalies in the Indian Ocean through advection.
- Early and deep formation of the ASL correlated with positive sea ice anomalies in the fall and winter in the Ross Sea and Bellingshausen and Amundsen Seas.
- Mid-winter formation of a weaker than normal zonal wave 3 pattern may have influenced late winter anomalies in the Ross Sea.

FUTURE NEEDS AND OPPORTUNITIES

This section draws from all of the workshop discussions to synthesize several recurrent and overarching themes, with particular attention to the needs for future observations and research. As in earlier sections, the themes presented here do not necessarily represent a consensus view among the participants, the view of the planning committee, or the view of the National Academies. Rather, these ideas are those of individuals or groups of workshop participants.

Observations

Improving the Satellite Record

Many workshop participants said the satellite record could be improved with the addition of the following satellite observations:

- Improved and validated satellite retrievals of large-scale sea ice edge and concentration, thickness, and snow cover and depth, including independent data for determining the

accuracy, bias, and precision of satellite sea ice concentration and extent algorithm and data (i.e., in marginal ice zone, covering seasons and regions);
- Ice volume;
- Ice velocity;
- Ice mass balance;
- Precipitation and snow accumulation and loss; and
- Large-scale motion, dynamics, and deformation at high resolution.

Oceanographic Measurements

Oceanographic measurements were also identified by many participants as important for understanding Antarctic sea ice variability. For example:

- In situ ocean observations to assess vertical and horizontal distributions of heat and freshwater anomalies. Such observations could also be used to assess seasonal persistence, propagation or (re)emergence of ocean thermal and freshwater anomalies.
- In situ ice and ocean observations to assess controls on seasonal ice and snow thickness evolution, interactions between the ocean and ice shelves, and regions of active upwelling (e.g., shelf break, canyon heads).
- Deep ocean measurements to assess temperature and salinity variability related to changes in sea ice and the impact of melt water at different depths.

Several participants said that new technology (e.g., autonomous surface vessels, gliders, tagged marine mammals, and cables) would be useful for making key oceanographic measurements such as

- Full-depth hydrographic profiles under ice and in winter (e.g., temperature and salinity);
- Lateral ocean circulation changes;
- Vertical heat flux as a function of time to better understand overturning circulation;
- Position of Antarctic Circumpolar Current fronts to determine if they move over time;
- Ice production and melt in polynyas and elsewhere;
- Stratification and mixed layer depth everywhere, as a function of time;
- Geochemical tracers for decomposing ocean freshwater budget;
- Four-dimensional ocean circulation fields, including time-varying overturning; and
- Vertical and isopycnal mixing changes.

Generating Longer Records of Sea Ice Variability

Several participants emphasized the importance of generating longer records of sea ice variability. Satellite data give good coverage of sea ice variability (e.g., extent, concentration, and area) and trends in recent era; however, it is necessary to reconstruct data before 1979 in order to better understand recent trends and help with attribution. Extending the historical record would also facilitate comparison with models to understand recent trends and the future. Some participants said that obtaining data from even just a few years can make a significant difference in understanding the system. Several participants also highlighted the benefits of using complementary observations (e.g., SST, MSLP) together with sparse sea ice extent observations. Options for extending the record before 1979 include

- Homogenization of ESMR data with post-1979 record,
- Gap filling with ice charts,
- Calibration of Nimbus record (e.g., 1964 and 1966 anomalies),
- Recalibration of the present data (post-2000) with scatterometer data against passive microwave data to better define sea ice extent (problems with 15% contour),
- Ice core chemistry (MSA record; for limited regions),
- Sediment records,
- Early ship/whaling records, and
- Penguin colony records.

Coupled Reanalysis Products

Several participants expressed significant concerns about reanalysis, especially given that it limits assessment of the sea ice trend and the role of westerly winds. It is difficult to assess the uncertainty of reanalysis and construct a homogenous record. Observations of ice thickness, depth of snow on top of ice, trends of ice drift, snow accumulation rates and patterns, precipitation, and clouds (e.g., shortwave) could help to constrain the reanalyses. Other participants suggested choosing specific problems and/or regions (e.g., Ross Sea) that are highlighted by models and use proxy data to test and validate the reanalysis. However, there was some skepticism that the proxy data could be trusted enough for this kind of validation. Many participants said there is promise for coupled reanalysis products, which use physics and other observations that go back further in time. Such products would be the only independent check to look back before 1970. They would also serve as another tool to compare with CMIP-level models.

Observations for Model Assessment

Many participants noted that observations are critical to support model assessment. The Southern Ocean State Estimate could be a good dataset for model evaluation; it has shown surface fluxes are critical for correctly estimating water masses. There has also been a push to do more projections and large ensembles rather than emphasizing finer scales to resolve eddies, polynyas, and katabatic winds. Other observations for improving models include surface energy balance and ocean temperature and salinity profiles under the ice.

Models

Improvements in Model Parameterizations

Several participants said that enhancements in model parameterizations are needed to improve model representation of sea ice trends and variability. Examples of parameterizations that need improvements include polynyas, ice drift (in particular, ice drag coefficients), mixed layer schemes under the sea ice, the role of clouds (especially given that the Southern Ocean is among the cloudiest[12] places on Earth; e.g., McCoy et al., 2015), and wave–sea ice interaction.

Modeling Studies and Intercomparisons

Workshop participants also discussed modeling studies and intercomparisons to advance understanding of Antarctic sea ice variability. Some participants said that there might be a value in doing experiments with a high-resolution atmospheric model, especially in the Ross Sea. Others suggested research to test how the sea ice response differs if the model is forced (nudged) toward the observed mean state would be useful. Another idea is to analyze a process-based budget to look at interannual variability (e.g., thermodynamics versus transport).

Because it is not always clear whether the trends are captured for the correct reasons, modeling studies that diagnose whether sea ice trends are captured for the right or wrong process reasons are critical, said several participants. This requires developing a set of hypotheses to test model sensitivities to certain processes. Then, focused process studies could be conducted to test these hypotheses.

Process-Based Understanding

Many participants said that process-based understanding is critical for improving understanding of the mechanisms of Antarctic sea ice variability. Process studies also provide validation of global coupled models. Such process studies could be comprised of drifting stations, could take into account the seasonal cycle, and could be supplemented by autonomous devices and high-resolution passive microwave observations (e.g., of marginal ice zone, ice concentration, and floe size). This approach was successfully executed in autumn 2015 in the Arctic on the Office of Naval Research SeaState campaign and is planned for the Ross Sea in autumn 2017. Examples of key processes to address in any Antarctic process study include understanding the roles of

- Spring/summer heat and freshwater anomalies in impacting subsequent ice-edge advance and growth processes;
- Winds in amplifying or dampening the seasonal ice-ocean feedbacks;

[12] Percentage of cloud cover year round.

- Ocean stratification in affecting these processes and thus the seasonal evolution of sea ice thickness;
- Concurrent changes in snow on sea ice in amplifying or dampening the response;
- Extreme environmental events in amplifying the impact or tipping the sea ice response;
- The marginal ice zone, interior, polynyas/coastal (e.g., polynya ice production rates) dynamics versus thermodynamics; and
- Waves in sea ice formation and melt.

Several participants also emphasized the importance of understanding the relationship between sea ice and shelf ice, in particular the role of sea ice in ocean–ice shelf interactions—for example, whether freshwater affects ice growth at the surface. It is also important to assess the response of sea ice to changes in coastal icescapes (polynyas, fast ice, glacier tongues, and iceberg fields), according to some participants.

Many workshop participants identified several regions of particular geographic interest for further study, particularly as part of a process study. The western Ross Sea is particularly anomalous, and scientists have not been able to identify any drivers for the trends there. The optimal time for studying this region is during sea ice advance. The Weddell Sea in summer is another outlier and candidate for further study.

Communicating to the Public

One theme that emerged from the discussion is the importance of communicating Antarctic sea ice trends to the public. Sea ice is an oft-referred symbol in public conversations about the evidence for human-caused climate warming. While the widely cited decline in Arctic sea ice is a powerful example of emerging climate change impacts, the modest sea ice increase in Antarctica, along with the record-breaking maximum years of 2012-2014, are occasionally and erroneously cited as evidence that human-caused climate change is not happening (Perovich, 2011).

Some participants said a schematic is needed to compare and contrast Arctic and Antarctic sea ice. Such a schematic could highlight the Antarctic's large overlying atmospheric variability compared to the Arctic, the different geographies of the two regions, and the different ocean circulation and heat exchange processes. Dr. Katy Barnhart, Annenberg Public Policy Center, University of Pennsylvania, said that she is conducting research to assist the climate science community with communicating to the public that Antarctic sea ice increase does not

- demonstrate that climate change is not happening,
- indicate that concern about Arctic sea ice decline is unwarranted, or
- mean that climate models are not to be trusted.

The lack of a ready explanation of Antarctic sea ice variability and its drivers, as well as discrepancies between models and observations, contributes to a lingering public misconception that recent trends in Antarctic sea ice disprove human-caused climate warming, or that contrasting trends in sea ice at the two poles somehow cancel each other out.

Given the importance of sea ice in the Earth system, many participants said that it is critical to advance understanding of the sea ice variability and change in the Southern Ocean surrounding Antarctica. Key observations, model impovements, and new research that were discussed during the workshop might clarify the processes and mechanisms that control observed sea ice variability and improve prediction of Antarctic sea ice.

References

Abram, N. J., E. R. Thomas, J. R. McConnell, R. Mulvaney, T. J. Bracegirdle, L. C. Sime, and A. J. Aristarain. 2010. Ice core evidence for a 20th century decline of sea ice in the Bellingshausen Sea, Antarctica. *Journal of Geophysical Research* 115:D23101, doi: 10.1029/2010JD014644.

Abram, N. J., E. W. Wolff, and M. A. J. Curran. 2013. A review of sea ice proxy information from polar ice cores. *Quaternary Science Reviews* 79:168-183, doi: 10.1016/j.quascirev.2013.01.011.

Ackley, S., P. Wadhams, J. C. Comiso, and A. P. Worby. 2003. Decadal decrease of Antarctic sea ice extent inferred from whaling records revisited on the basis of historical and modern sea ice records. *Polar Research* 22(1):19-25, doi: 10.1111/j.1751-8369.2003.tb00091.x.

Armour, K. C., and C. M. Bitz. 2015. Observed and projected trends in Antarctic sea ice. *US Clivar Variations* 13(4):12-19.

Armour, K. C., J. Marshall, J. R. Scott, A. Donohoe, and E. R. Newsom. 2016. Southern Ocean warming delayed by circumpolar upwelling and equatorward transport. *Nature Geoscience* 9:549-554, doi: 10.1038/ngeo2731.

Arrigo, K. R., M. P. Lizotte, and T. Mock. 2009. Sea ice algae. In *Sea Ice*, D. N. Thomas and G. S. Dieckmann, eds. Oxford, UK: Blackwell Science, Ltd.

Bintanja, R., G. J. van Oldenborgh, S. S. Drijfhout, B. Wouters, and C. A. Katsman. 2013. Important role for ocean warming and increased ice-shelf melt in Antarctic sea ice expansion. *Nature Geoscience* 6(5):376-379, doi: 10.1038/ngeo1767.

Bitz, C., and L. Polvani. 2012. Antarctic climate response to stratospheric ozone depletion in a fine resolution ocean climate model. *Geophysical Research Letters* 39(20), doi: 10.1029/2012GL053393.

Bitz, C. M., P. R. Gent, R. A. Woodgate, M. M. Holland, and R. Lindsay. 2006. The influence of sea ice on ocean heat uptake in response to increasing CO_2. *Journal of Climate* 19(11):2437-2450, doi: 10.1175/JCLI3756.1.

Böning, C. W., A. Dispert, M. Visbeck, S. R. Rintoul, and F. U. Schwarzkopf. 2008. The response of the Antarctic Circumpolar Current to recent climate change. *Nature Geoscience* 1(12):864-869, doi: 10.1038/ngeo362.

Brucker, L., E. P. Dinnat, G. Picard, and N. Champollion. 2014. Effect of snow surface metamorphism on Aquarius L-band radiometer observations at Dome C, Antarctica. *IEEE Transactions on Geoscience and Remote Sensing* 52(11):7408-7417, doi: 10.1109/TGRS.2014.2312102.

Cavalieri, D. J., C. L. Parkinson, and K. Y. Vinnikov. 2003. 30-year satellite record reveals contrasting Arctic and Antarctic decadal sea ice variability. *Geophysical Research Letters* 30(18), doi: 10.1029/2003GL018031.

Ciasto, L. M., and D. W. J. Thompson. 2008. Observations of large-scale ocean-atmosphere interaction in the Southern Hemisphere. *Journal of Climate* 21(6):1244-1259, doi: 10.1175/2007JCLI1809.1.

Ciasto, L. M., G. R. Simpkins, and M. H. England. 2015. Teleconnections between tropical Pacific SST anomalies and extratropical Southern Hemisphere climate. *Journal of Climate* 28(1):56-65, doi: 10.1175/JCLI-D-14-00438.1.

Comiso, J. C., and F. Nishio. 2008. Trends in the sea ice cover using enhanced and compatible AMSR-E, SSM/I, and SMMR data. *Journal of Geophysical Research: Oceans* 113(C2), doi: 10.1029/2007JC004257.

Comiso, J. C., D. J. Cavalieri, C. L. Parkinson, and P. Gloersen. 1997. Passive microwave algorithms for sea ice concentration: A comparison of two techniques. *Remote Sensing of Environment* 60(3):357-384, doi: 10.1016/S0034-4257(96)00220-9.

Comiso, J. C., D. J. Cavalieri, and T. Markus. 2003. Sea ice concentration, ice temperature, and snow depth using AMSR-E data. *IEEE Transactions on Geoscience and Remote Sensing* 41(2):243-252, doi: 10.1109/Tgrs.2002.808317.

Comiso, J. C., R. Kwok, S. Martin, and A. L. Gordon. 2011. Variability and trends in sea ice extent and ice production in the Ross Sea. *Journal of Geophysical Research: Oceans* 116(C4), doi: 10.1029/2010JC006391.

Cotté, C., and C. Guinet. 2007. Historical whaling records reveal major regional retreat of Antarctic sea ice. *Deep Sea Research Part I: Oceanographic Research Papers* 54(2):243-252, doi: 10.1016/j.dsr.2006.11.001.

Curran, M. A. J., T. D. van Ommen, V. I. Morgan, K. L. Phillips, and A. S. Palmer. 2003. Ice core evidence for Antarctic sea ice decline since the 1950s. *Science* 302(564):1203-1206, doi: 10.1126/science.1087888.

Curran, M. A. J., T. D. Van Ommen, A. Moy, and C. Plummer. 2010. *High resolution climate proxy records from Australian Antarctic ice core research*. Australia New Zealand Climate Forum. Conference Paper. October 13-15, 2010, Hobart, Australia.

Cvijanovic, I., and K. Caldeira. 2015. Atmospheric impacts of sea ice decline in CO_2 induced global warming. *Climate Dynamics* 44(5-6):1173-1186, doi: 10.1007/s00382-015-2489-1.

de la Mare, W. K. 1997. Abrupt mid-twentieth-century decline in Antarctic sea-ice extent from whaling records. *Nature* 389(6646):57-60, doi: 10.1038/37956.

de la Mare, W. K. 2009. Changes in Antarctic sea-ice extent from direct historical observations and whaling records. *Climatic Change* 92(3-4):461-493, doi: 10.1007/s10584-008-9473-2.

Deser, C., A. Phillips, V. Bourdette, and H. Teng. 2012. Uncertainty in climate change projections: The role of internal variability. *Climate Dynamics* 38(3-4):527-546, doi: 10.1007/s00382-010-0977-x.

Ding, Q., and E. J. Steig. 2013. Temperature change on the Antarctic Peninsula linked to the Tropical Pacific. *Journal of Climate* 26(19):7570-7585, doi: 10.1175/JCLI-D-12-00729.1.

Ding, Q., E. J. Steig, D. S. Battisti, and M. Küttel. 2011. Winter warming in West Antarctica caused by central tropical Pacific warming. *Nature Geoscience* 4(6):398-403, doi: 10.1038/ngeo1129.

Ding, Q., E. J. Steig, D. S. Battisti, and J. M. Wallace. 2012. Influence of the tropics on the Southern Annular Mode. *Journal of Climate* 25(18):6330-6348, doi: 10.1175/JCLI-D-11-00523.1.

Douville, H., A. Voldoire, and O. Geoffroy. 2015. The recent global warming hiatus: What is the role of Pacific variability? *Geophysical Research Letters* 42(3):880-888, doi: 10.1002/2014GL062775.

Eisenman, I., W. N. Meier, and J. R. Norris. 2014. A spurious jump in the satellite record: Has Antarctic sea ice expansion been overestimated? *The Cryosphere* 8(4):1289-1296, doi: 10.5194/tc-8-1289-2014.

Fan, T., C. Deser, and D. P. Schneider. 2014. Recent Antarctic sea ice trends in the context of Southern Ocean surface climate variations since 1950. *Geophysical Research Letters* 41(7):2419-2426, doi: 10.1002/2014GL059239.

Ferreira, D., J. Marshall, C. M. Bitz, S. Solomon, and A. Plumb. 2015. Antarctic Ocean and sea ice response to ozone depletion: A two-time-scale problem. *Journal of Climate* 28(3):1206-1226, doi: 10.1175/JCLI-D-14-00313.1.

Fetterer, F., K. Knowles, W. Meier, and M. Savoie. 2016, updated daily. *Sea Ice Index, Version 2*. National Snow and Ice Data Center. doi: 10.7265/N5736NV7.

Fogt, R. L., and D. H. Bromwich. 2006. Decadal variability of the ENSO teleconnection to the high-latitude South Pacific governed by coupling with the Southern Annular Mode. *Journal of Climate* 19(6):979-997, doi: 10.1175/JCLI3671.1.

Fogt, R. L., and T. A. Scambos, Eds. 2013. Antarctica [in State of the Climate in 2012]. *Bulletin of the American Meteorological Society* 94(8):S133-S147.

Fogt, R. L., and T. A. Scambos, Eds. 2014. Antarctica [in State of the Climate in 2013]. *Bulletin of the American Meteorological Society* 95(7):S143-S157.

Fogt, R. L., and A. J. Wovrosh. 2015. The relative influence of tropical sea surface temperatures and radiative forcing on the Amundsen Sea Low. *Journal of Climate* 28(21):8540-8555, doi: 10.1175/JCLI-D-15-0091.1.

Fogt, R. L., and E.A. Zbacnik. 2014. Sensitivity of the Amundsen Sea Low to stratospheric ozone depletion. *Journal of Climate* 27(24):9383-9400, doi: 10.1175/JCLI-D-13-00657.1.

Fogt, R. L., D. H. Bromwich, and K. M. Hines. 2011. Understanding the SAM influence on the South Pacific ENSO teleconnection. *Climate Dynamics* 36(7-8):1555-1576, doi: 10.1007/s00382-010-0905-0.

Fraser, A. D., R. A. Massom, K. J. Michael, B. K. Galton-Fenzi, and J. L. Lieser. 2012. East Antarctic landfast sea ice distribution and variability, 2000-08. *Journal of Climate* 25(4):1137-1156, doi: 10.1175/JCLI-D-10-05032.1.

Frölicher, T. L., J. L. Sarmiento, D. J. Paynter, J. P. Dunne, J. P. Krasting, and M. Winton. 2015. Dominance of the Southern Ocean in anthropogenic carbon and heat uptake in CMIP5 models. *Journal of Climate* 28(2):862-886, doi: 10.1175/JCLI-D-14-00117.1.

Gagné, M.-È., N. P. Gillett, and J. C. Fyfe. 2015. Impact of aerosol emission controls on future Arctic sea ice cover. *Geophysical Research Letters* 42(20):8481-8488, doi: 10.1002/2015GL065504.

Gallaher, D. W., G. G. Campbell, and W. N. Meier. 2014. Anomalous variability in Antarctic sea ice extents during the 1960s with the use of Nimbus data. *IEEE Journal of Selected Topics in Applied Earth Observations and Remote Sensing* 7(3):881-887, doi: 10.1109/JSTARS.2013.2264391.

Giles, A. B., R. A. Massom, and V. I. Lytle. 2008. Fast-ice distribution in East Antarctica during 1997 and 1999 determined using RADARSAT data. *Journal of Geophysical Research: Oceans* 113(C2), doi: 10.1029/2007JC004139.

Goosse, H., and V. Zunz. 2014. Decadal trends in the Antarctic sea ice extent ultimately controlled by ice-ocean feedback, *The Cryosphere* 8(2):453-470, doi: 10.5194/tc-8-453-2014.

Goosse, H., W. Lefebvre, A. de Montety, E. Crespin, and A. Orsi. 2009. Consistent past half century trends in the atmosphere, the sea ice and the ocean at high southern latitudes. *Climate Dynamics* 33:999-1016, doi: 10.1007/s00382-008-0500-9.

Halfar, J., W. H. Adey, A. Kronz, S. Hetzinger, E. Edinger, and W. W. Fitzhugh. 2013 . Arctic sea-ice decline archived by multicentury annual-resolution record from crustose coralline algal proxy. *Proceedings of the National Academy of Sciences* 110(49):19737-19741, doi: 10.1073/pnas.1313775110.

Haumann, F. A. 2011. Dynamical interaction between atmosphere and sea ice in Antarctica. MSc Thesis. Utrecht, Netherlands: Utrecht University.

Haumann, F. A., D. Notz, and H. Schmidt. 2014. Anthropogenic influence on recent circulation driven Antarctic sea ice changes. *Geophysical Research Letters* 41:8429-8437, doi: 10.1002/2014GL061659.

Hellmer, H. H. 2004. Impact of Antarctic ice shelf basal melting on sea ice and deep ocean properties. *Geophysical Research Letters* 31(10), doi: 10.1029/2004gl019506.

Holland, P. R. 2014. The seasonality of Antarctic sea ice trends. *Geophysical Research Letters* 41(12):4230-4237, doi: 10.1002/2014GL060172.

Holland, P. R., and R. Kwok. 2012. Wind-driven trends in Antarctic sea-ice drift. *Nature Geoscience* 5(12):872-875, doi: 10.1038/ngeo1627.

Holland, P. R., N. Bruneau, C. Enright, M. Losch, N. T. Kurtz, and R. Kwok. 2014. Modeled trends in Antarctic sea ice thickness. *Journal of Climate* 27(10):3784-3801, doi: 10.1175/JCLI-D-13-00301.1.

IPCC (Intergovernmental Panel on Climate Change). 2013. *Climate Change 2013: The Physical Science Basis.* Contribution of Working Group I to the Fifth Assessment Report of the Intergovernmental Panel on Climate Change. T. F. Stocker, D. Qin, G.-K. Plattner, M. Tignor, S. K. Allen, J. Boschung, A. Nauels, Y. Xia, V. Bex and P. M. Midgley, eds. Cambridge, UK: Cambridge University Press.

Karoly, D. J. 1989. Southern Hemisphere circulation features associated with El Niño-Southern Oscillation events. *Journal of Climate* 2(11):1239-1252, doi: 10.1175/1520-0442(1989)002<1239:SHCFAW>2.0.CO;2.

Khazendar, A., M. P. Schodlok, I. Fenty, S. R. M. Ligtenberg, E. Rignot, and M. R. Van den Broeke. 2013. Observed thinning of Totten Glacier is linked to coastal polynya variability. *Nature Communications* 4, doi: 10.1038/ncomms3857.

Kirkman IV, C. H., and C. M. Bitz. 2011. The effect of the sea ice freshwater flux on Southern Ocean temperatures in CCSM3: Deep-ocean warming and delayed surface warming. *Journal of Climate* 24(9):2224-2237, doi: 10.1175/2010JCLI3625.1.

Kosaka, Y., and S. P. Xie. 2013. Recent global-warming hiatus tied to equatorial Pacific surface cooling. *Nature* 501(7467):403-407, doi: 10.1038/nature12534.

Kostov, Y., J. Marshall, U. Hausmann, K. C. Armour, D. Ferreira, and M. M. Holland. 2016. Fast and slow responses of Southern Ocean sea surface temperature to SAM in coupled climate models. *Climate Dynamics* 1-15, doi: 10.1007/s00382-016-3162-z.

Kurtz, N. T., and T. Markus. 2012. Satellite observations of Antarctic sea ice thickness and volume. *Journal of Geophysical Research: Oceans* 117(C8), doi: 10.1029/2012JC008141.

Lecomte, O., H. Goosse, T. Fichefet, P. R. Holland, P. Uotila, V. Zunz, and N. Kimura. 2016. Impact of surface wind biases on the Antarctic sea ice concentration budget in climate models. *Ocean Modelling* 105:60-70, doi: 10.1016/j.ocemod.2016.08.001.

Lefebvre, W., H. Goosse, R. Timmermann, and T. Fichefet. 2004. Influence of the Southern Annular Mode on the sea ice–ocean system. *Journal of Geophysical Research: Oceans* 109(C9), doi: 10.1029/2004JC002403.

Leonard, K. C., and R. I. Cullather. 2008. Snowfall measurements in the Amundsen and Bellingshausen Seas, Antarctica. *Proceedings of the Eastern Snow Conference* 65:87-98.

Leonard, K. C., and T. Maksym. 2011. The importance of wind-blown snow redistribution to snow accumulation on Bellingshausen Sea ice. *Annals of Glaciology* 52(57):271-278.

L'Heureux, M. L., and D. W. J. Thompson. 2006. Observed relationships between the El Niño-Southern Oscillation and the extratropical zonal-mean circulation. *Journal of Climate* 19(2):276-287, doi: 10.1175/JCLI3617.1.

Li, X., D. M. Holland, E. P. Gerber, and C. Yoo. 2014. Impacts of the north and tropical Atlantic Ocean on the Antarctic Peninsula and sea ice. *Nature* 505(7484):538-542, doi: 10.1038/nature12945.

Liu, J., and J. A. Curry. 2010. Accelerated warming of the Southern Ocean and its impacts on the hydrological cycle and sea ice. *Proceedings of the National Academy of Sciences* 107(34):14987-14992, doi: 10.1073/pnas.1003336107.

Liu, J. P., X. J. Yuan, D. G. Martinson, and D. Rind. 2004. Re-evaluating Antarctic sea-ice variability and its teleconnections in a GISS global climate model with improved sea ice and ocean processes. *International Journal of Climatology* 24:841-852, doi: 10.1002/joc.1040.

Mackintosh, N. A. 1972. Sea ice limits. *Discovery Reports* 36(13).

Mahlstein, I., P. R. Gent, and S. Solomon. 2013. Historical Antarctic mean sea ice area, sea ice trends, and winds in CMIP5 simulations. *Journal of Geophysical Research: Atmospheres* 118(11):5105-5110, doi: 10.1002/jgrd.50443.

Maksym, T., and T. Markus. 2008. Antarctic sea ice thickness and snow-to-ice conversion from atmospheric reanalysis and passive microwave snow depth. *Journal of Geophysical Research: Oceans* 113(C2), doi: 10.1029/2006jc004085.

Maksym, T., S. E. Stammerjohn, S. Ackley, and R. Massom. 2012. Antarctic sea ice: A polar opposite? *Oceanography* 25(3):140-151, doi: 10.5670/oceanog.2012.88.

Manabe, S., R. J. Stouffer, M. J. Spelman, and K. Bryan. 1991. Transient responses of a coupled ocean-atmosphere model to gradual changes of atmospheric CO_2. Part I. Annual mean response. *Journal of Climate* 4(8):785-818, doi: 10.1175/1520-0442(1991)004<0785:TROACO>2.0.CO;2.

Marshall, J., K. C. Armour, J. R. Scott, Y. Kostov, U. Hausmann, D. Ferreira, T. G. Shepherd, and C. M. Bitz. 2014. The ocean's role in polar climate change: Asymmetric Arctic and Antarctic responses to greenhouse gas and ozone forcing. *Philosophical Transactions of the Royal Society of London A: Mathematical, Physical and Engineering Sciences* 372(2019):20130040, doi: 10.1098/rsta.2013.0040.

Martinson, D. G., and R. A. Iannuzzi. 1998. Antarctic ocean-ice interaction: Implications from ocean bulk property distributions in the Weddell Gyre. In *Antarctic Research Series, Antarctic Sea Ice: Physical Processes, Interactions and Variability*, M. Jeffries, ed. Boston: American Geophysical Union, doi: 10.1029/AR074p0243.

Massom, R. A., and S. E. Stammerjohn. 2010. Antarctic sea ice change and variability—physical and ecological implications. *Polar Science* 4(2):149-186, doi: 10.1016/j.polar.2010.05.001.

Massom, R. A., H. Eicken, C. Hass, M. O. Jeffries, M. R. Drinkwater, M. Sturm, A. P. Worby, X. Wu, V. I. Lytle, S. Ushio, K. Morris, P. A. Reid, S. G. Warren, and I. Allison. 2001. Snow on Antarctic sea ice. *Reviews of Geophysics* 39(3):413-445, doi: 10.1029/2000RG000085.

Massom, R. A., S. E. Stammerjohn, R. C. Smith, M. J. Pook, R. A. Iannuzzi, N. Adams, D. G. Martinson, M. Vernet, W. R. Fraser, L. B. Quetin, R. M. Ross, Y. Massom, and H. R. Krouse. 2006. Extreme anomalous atmospheric circulation in the West Antarctic Peninsula region in austral spring and summer 2001/2, and its profound impact on sea ice and biota. *Journal of Climate* 19:3544-3571.

Massom, R. A., A. B. Giles, H. A. Fricker, R. C. Warner, B. Legresy, G. Hyland, N. Young, and A. D. Fraser. 2010. Examining the interaction between multi-year landfast sea-ice and the Mertz Glacier tongue, East Antarctica: Another factor in ice sheet stability? *Journal of Geophysical Research* 115:C12027, doi: 10.1029/2009JC006083.

Massonnet, F., V. Guemas, N. S. Fuckar, and F. J. Doblas-Reyes. 2015. The 2014 high record of Antarctic sea ice extent [in "Explaining Extremes of 2014 from a Climate Perspective"]. *Bulletin of the American Meteorological Society* 96(12):S163-S167.

Matear, R. J., T. J. O'Kane, J. S. Risbey, and M. Chamberlain. 2015. Sources of heterogeneous variability and trends in Antarctic sea-ice. *Nature Communications* 6, doi: 10.1038/ncomms9656.

McCoy, D. T., S. M. Burrows, R. Wood, D. P. Grosvenor, S. M. Elliott, P.-L. Ma, P. J. Rasch, and D. L. Hartmann. 2015. Natural aerosols explain seasonal and spatial patterns of Southern Ocean cloud albedo. *Science Advances* 1(6):e1500157, doi: 10.1126/sciadv.1500157.

Meehl, G. A., A. Hu, and H. Teng. 2016. Initialized decadal prediction for transition to positive phase of the Interdecadal Pacific Oscillation. *Nature Communications* 7, doi: 10.1038/ncomms11718.

Meier, W. N., S. Gerland, M. A. Granskog, J. R. Key, C. Haas, G. K. Hovelsrud, K. Kovacs, A. Makshtas, C. Michel, D. Perovich, J. D. Reist, and B. E. h. van Oort. 2011. Chapter 9: Sea ice. In *Snow, Water, Ice and Permafrost in the Arctic (SWIPA): Climate Change and the Cryosphere.* Oslo, Norway: Arctic Monitoring and Assessment Programme.

Meier, W. N., D. Gallaher, and G. G. Campbell. 2013. New estimates of Arctic and Antarctic sea ice extent during September 1964 from recovered Nimbus I satellite imagery. *The Cryosphere* 7(2):699-705, doi: 10.5194/tc-7-699-2013.

Meredith, M. P., H. J. Venables, A. Clarke, H. W. Ducklow, M. Erickson, M. J. Leng, J. T. M. Lenaerts, and M. R. van den Broeke. 2013. The freshwater system west of the Antarctic Peninsula: Spatial and temporal changes. *Journal of Climate* 26(5):1669-1684, doi: 10.1175/JCLI-D-12-00246.1.

Miles, B. W. J., C. R. Stokes, and S. S. R. Jamieson. 2016. Pan-ice-sheet glacier terminus change in East Antarctica reveals sensitivity of Wilkes Land to sea-ice changes. *Science Advances* 2(5):e1501350, doi: 10.1126/sciadv.1501350.

Mo, K. C. 2000. Relationships between low-frequency variability in the Southern Hemisphere and sea surface temperature anomalies. *Journal of Climate* 13(20):3599-3610, doi: 10.1175/1520-0442(2000)013<3599:RBLFVI>2.0.CO;2.

Mo, K. C., and R. W. Higgins. 1998. The Pacific-South American modes and tropical convection during the Southern Hemisphere winter. *Monthly Weather Review* 126(6):1581-1596, doi: 10.1175/1520-0493(1998)126<1581:TPSAMA>2.0.CO;2.

Murphy, E. J., A. Clarke, C. Symon, and J. Priddle. 1995. Temporal variation in Antarctic sea ice: Analysis of a long term fast-ice record from the South Orkney Islands. *Deep Sea Research Part I: Oceanographic Research Papers* 42(7):1045-1062, doi: 10.1016/0967-0637(95)00057-D.

Murphy, E. J., A. Clarke, N. J. Abram, and J. Turner. 2014. Variability of seaice in the northern Weddell Sea during the 20th century. *Journal of Geophysical Research Oceans* 119:4549-4572, doi: 10.1002/2013JC009511.

Parkinson, C. L. 2014. Global sea ice coverage from satellite data: Annual cycle and 35-yr trends. *Journal of Climate* 27(24):9377-9382, doi: 10.1175/JCLI-D-14-00605.1.

Parkinson, C. L., and D. J. Cavalieri. 2012. Antarctic sea ice variability and trends, 1979–2010. *The Cryosphere* 6(4):871-880, doi: 10.5194/tc-6-871-2012.

Parkinson, C. L., and J. C. Comiso. 2008. Antarctic sea ice parameters from AMSR-E data using two techniques and comparisons with sea ice from SSM/I. *Journal of Geophysical Research* 113:(C02S06), doi: 10.1029/2007JC004253.

Parkinson, C. L., and J. C. Comiso. 2013. On the 2012 record low Arctic sea ice cover: Combined impact of preconditioning and an August storm. *Geophysical Research Letters* 40(7):1356-1361, doi: 10.1002/grl.50349.

Pauling, A. G., C. M. Bitz, I. J. Smith, and P. J. Langhorne. 2016. The response of the Southern Ocean and Antarctic sea ice to freshwater from ice shelves in an Earth system model. *Journal of Climate* 29(5):1655-1672, doi: 10.1175/JCLI-D-15-0501.1.

Perovich, D. K. 2011. The changing Arctic sea ice cover. *Oceanography* 24(3):162-173, doi: 10.5670/oceanog.2011.68.

Polvani, L. M., and K. L. Smith. 2013. Can natural variability explain observed Antarctic sea ice trends? New modeling evidence from CMIP5. *Geophysical Research Letters* 40(12):3195-3199, doi: 10.1002/grl.50578.

Polvani, L. M., D. W. Waugh, G. J. P. Correa, and S. W. Son. 2011. Stratospheric ozone depletion: The main driver of twentieth-century atmospheric circulation changes in the Southern Hemisphere. *Journal of Climate* 24(3):795-812, doi: 10.1175/2010JCLI3772.1.

Raphael, M. N. 2007. The influence of atmospheric zonal wave three on Antarctic sea ice variability. *Journal of Geophysical Research: Atmospheres* 112(D12), doi: 10.1029/2006JD007852.

Raphael, M. N., and W. Hobbs. 2014. The influence of the large-scale atmospheric circulation on Antarctic sea ice during ice advance and retreat seasons. *Geophysical Research Letters* 41(14):5037-5045, doi: 10.1002/2014GL060365.

Raphael, M. N., G. J. Marshall, J. Turner, R. L. Fogt, D. Schneider, D. A. Dixon, J. S. Hosking, J. M. Jones, and W. R. Hobbs. 2016. The Amundsen Sea Low: Variability, change, and impact on Antarctic climate. *Bulletin of the American Meteorological Society* 97(1):111-121, doi: 10.1175/BAMS-D-14-00018.1.

Reid, P., and R. Massom. 2015. Sidebar 6.3: Successive Antarctic sea ice extent records during 2012, 2013, and 2014 [in "State of the Climate in 2014"]. *Bulletin of the American Meteorological Society* 96(7): S163-S164.

Reid, P., S. Stammerjohn, R. A. Massom, T. Scambos, and J. Lieser. 2015. The record 2013 Southern Hemisphere sea-ice extent maximum. *Annals of Glaciolology* 56:99-106, doi: 10.3189/2015AoG69A892.

Rhein, M., S. R. Rintoul, S. Aoki, E. Campos, D. Chambers, R. A. Feely, S. Gulev, G. C. Johnson, S. A. Josey, A. Kostianoy, C. Mauritzen, D. Roemmich, L. D. Talley, and F. Wang. 2013. Observations: Ocean. In: *Climate Change 2013: The Physical Science Basis.* Contribution of Working Group I to the Fifth Assessment Report of the Intergovernmental Panel on Climate Change. Cambridge, UK: Cambridge University Press.

Rye, C. D., A. C. N. Garabato, P. R. Holland, M. P. Meredith, A. J. G. Nurser, C. W. Hughes, A. C. Coward, and D. J. Webb. 2014. Rapid sea-level rise along the Antarctic margins in response to increased glacial discharge. *Nature Geoscience* 7(10):732-735, doi: 10.1038/ngeo2230.

Schneider, D. P., Y. Okumura, and C. Deser. 2012a. Observed Antarctic interannual climate variability and tropical linkages. *Journal of Climate* 25(12):4048-4066, doi: 10.1175/JCLI-D-11-00273.1.

Schneider, D. P., C. Deser, and Y. Okumura. 2012b. An assessment and interpretation of the observed warming of West Antarctica in the austral spring. *Climate Dynamics* 38(1):323-347, doi: 10.1007/s00382-010-0985-x.

Screen, J. A., and I. Simmonds. 2010. The central role of diminishing sea ice in recent Arctic temperature amplification. *Nature* 464(7293):1334-1337, doi: 10.1038/nature09051.

Sen Gupta, A., and M. H. England. 2006. Coupled ocean-atmosphere-ice response to variations in the Southern Annular Mode. *Journal of Climate* 19(18):4457-4486, doi: 10.1175/JCLI3843.1.

Sigmond, M., and J. C. Fyfe. 2010. Has the ozone hole contributed to increased Antarctic sea ice extent? *Geophysical Research Letters* 37(18), doi: 10.1029/2010GL044301.

Simpkins, G. R., L. M. Ciasto, D. W. J. Thompson, and M. H. England. 2012. Seasonal relationships between large-scale climate variability and Antarctic sea ice concentration. *Journal of Climate* 25(16):5451-5469, doi: 10.1175/JCLI-D-11-00367.1.

Simpkins, G. R., L. M. Ciasto, and M. H. England. 2013. Observed variations in multidecadal Antarctic sea ice trends during 1979–2012, *Geophysical Research Letters* 40:3643-3648, doi: 10.1002/grl.50715.

Simpkins, G. R., S. McGregor, A. S. Taschetto, L. M. Ciasto, and M. H. England. 2014. Tropical connections to climatic change in the extratropical Southern Hemisphere: The role of Atlantic SST trends. *Journal of Climate* 27(13):4923-4936, doi: 10.1175/JCLI-D-13-00615.1.

Sinclair, K. E., N. A. N. Bertler, M. M. Bowen, and K. R. Arrigo. 2014. Twentieth century sea ice trends in the Ross Sea from a high-resolution, coastal ice-core record. *Geophysical Research Letters* 41:3510–3516, doi: 10.1002/2014GL059821.

Sissala, J. E., R. R. Sabatini, and H. J. Ackermann. 1972. Nimbus satellite data for polar ice survey. *Polar Record* 16(102):367-373, doi: 10.1017/S0032247400063051.

Stammerjohn, S. E., D. G. Martinson, R. C. Smith, X. Yuan, and D. Rind. 2008. Trends in Antarctic annual sea ice retreat and advance and their relation to El Niño-Southern Oscillation and Southern Annular Mode variability. *Journal of Geophysical Research* 108:C03S90, doi: 10.1029/2007JC004269.

Stammerjohn, S., R. Massom, D. Rind, and D. Martinson. 2012. Regions of rapid sea ice change: An inter-hemispheric seasonal comparison. *Geophysical Research Letters* 39(6), doi: 10.1029/2012GL050874.

Streten, N. A. 1973. Some characteristics of satellite-observed bands of persistent cloudiness over the Southern Hemisphere. *Monthly Weather Review* 101(6):486-495, doi: 10.1175/1520-0493(1973)101<0486:SCOSBO>2.3.CO;2.

Stroeve, J. C., V. Kattsov, A. Barrett, M. Serreze, T. Pavlova, M. Holland, and W. N. Meier. 2012. Trends in Arctic sea ice extent from CMIP5, CMIP3 and observations. *Geophysical Research Letters* 39(16), doi: 10.1029/2012GL052676.

Swart, N. C., and J. C. Fyfe. 2012. Observed and simulated changes in the Southern Hemisphere surface westerly wind-stress. *Geophysical Research Letters* 39(16), doi: 10.1029/2012GL052810.

Swart, N. C., and J. C. Fyfe. 2013. The infuence of recent Antarctic ice sheet retreat on simulated sea ice area trends. *Geophysical Research Letters* 40(16):4328-4332, doi: 10.1002/grl.50820.

Thompson, D.W. J, and S. Solomon. 2002. Interpretation of recent Southern Hemisphere climate change. *Science* 296(5569):895-899, doi: 10.1126/science.1069270.

Thompson, D. W. J., S. Solomon, P. J. Kushner, M. H. England, K. M. Grise, and D. J. Karoly. 2011. Signatures of the Antarctic ozone hole in Southern Hemisphere surface climate change. *Nature Geoscience* 4:741-749, doi: 10.1038/ngeo1296.

Turner, J. 2004. The El Niño-Southern Oscillation and Antarctica. *International Journal of Climatology* 24(1):1-31, doi: 10.1002/joc.965.

Turner, J., J. C. Comiso, G. J. Marshall, T. A. Lachlan-Cope, T. Bracegirdle, T. Maksym, M. P. Meredith, and Z. Wang. 2009. Nonannular atmospheric circulation change induced by stratospheric ozone depletion and its role in the recent increase of Antarctic sea ice extent. *Geophysical Research Letters* 36(8), doi: 10.1029/2009GL037524.

Turner, J., J. S. Hosking, T. Phillips, and G. J. Marshall. 2013a. Temporal and spatial evolution of the Antarctic sea ice prior to the September 2012 record maximum extent. *Geophysical Research Letters* 40:5894-5898, doi: 10.1002/2013GL058371.

Turner, J., T. J. Bracegirdle, T. Phillips, G. J. Marshall, and J. S. Hosking. 2013b. An initial assessment of Antarctic sea ice extent in the CMIP5 models. *Journal of Climate* 26(5):1473-1484, doi: 10.1175/JCLI-D-12-00068.1.

Turner, J., J. S. Hosking, T. J. Bracegirdle, G. J. Marshall, and T. Phillips. 2015. Recent changes in Antarctic Sea ice. *Philosophical Transactions of the Royal Society of London A: Mathematical, Physical and Engineering Sciences* 373(2045):20140163, doi: 10.1098/rsta.2014.0163.

Turner, J., J. S. Hosking, G. J. Marshall, T. Phillips, and T. J. Bracegirdle. 2016. Antarctic sea ice increase consistent with intrinsic variability of the Amundsen Sea Low. *Climate Dynamics* 46(7-8):2391-2402, doi: 10.1007/s00382-015-2708-9.

Waugh, D. W., F. Primeau, T. DeVries, and M. Holzer. 2013. Recent changes in the ventilation of the southern oceans. *Science* 339(6119):568-570, doi: 10.1126/science.1225411.

Welch, K. A., P. A. Mayewski, and S. I. Whitlow. 1993. Methanesulfonic acid in coastal Antarctic snow related to sea-ice extent. *Geophysical Research Letters* 20(6):443-446, doi: 10.1029/93GL00499.

Worby, A. P., and J. C. Comiso. 2004. Studies of the Antarctic sea ice edge and ice extent from satellite and ship observations. *Remote Sensing of Environment* 92(1):98-111, doi: 10.1016/j.rse.2004.05.007.

Worby, A. P., T. Markus, A. D. Steer, V. I. Lytle, and R. A. Massom. 2008. Evaluation of AMSR-E snow depth product over East Antarctic sea ice using in situ measurements and aerial photography. *Journal of Geophysical Research: Oceans* 113(C5), doi: 10.1029/2007JC004181.

Yuan, X. 2004. ENSO-related impacts on Antarctic sea ice: A synthesis of phenomenon and mechanisms. *Antarctic Science* 16(4):415-425, doi: 10.1017/S0954102004002238.

Yuan, X., and C. Li. 2008. Climate modes in southern high latitudes and their impacts on Antarctic sea ice. *Journal of Geophysical Research: Oceans* 113(C6), doi: 10.1029/2006JC004067.

Yuan, X., and D. G. Martinson. 2000. Antarctic sea ice extent variability and its global connectivity. *Journal of Climate* 13(10):1697-1717, doi: 10.1175/1520-0442(2000)013<1697:ASIEVA>2.0.CO;2.

Zhang, J. 2007. Increasing Antarctic sea ice under warming atmospheric and oceanic conditions. *Journal of Climate* 20(11):2515-2529, doi: 10.1175/JCLI4136.1.

Zhang, J. 2014. Modeling the impact of wind intensification on Antarctic sea ice volume. *Journal of Climate* 27(1):202-214, doi: 10.1175/JCLI-D-12-00139.1.

Zunz, V., H. Goosse, and F. Massonnet, 2013. How does internal variability infuence the ability of CMIP5 models to reproduce the recent trend in Southern Ocean sea ice extent? *The Cryosphere* 7(2):451-468, doi: 10.5194/tc-7-451-2013.

Zwally, H. J., J. C. Comiso, C. L. Parkinson, W. J. Campbell, F. D. Carsey, and P. Gloersen. 1983. Antarctic sea ice, 1973-1976: Satellite passive-microwave observations. Report No. NASA-SP-459. Washington, DC: National Aeronautics and Space Administration.

Zwally, H. J., W. Abdalati, T. Herring, K. Larson, J. Saba, and K. Steffen. 2002. Surface melt induced acceleration of Greenland ice-sheet flow. *Science* 297(5579):218-222, doi: 10.1126/science.1072708.

Appendix A
Statement of Task

An ad hoc National Academies of Sciences, Engineering, and Medicine committee will plan a workshop to bring together Antarctic sea ice experts to review the current science for climate-ocean–ice interactions in the Southern Hemisphere, with an outlook back 50 years and an emphasis on the record extents observed in the last 3 years. Workshop attendees will be asked to examine the processes governing decadal-scale growth and recent surge in Antarctic sea ice extent and projections of future Antarctic sea ice changes, and identify how to improve our understanding of current and future Antarctic sea ice changes. Specific topics to be addressed at the workshop include

1. Observations of changes in Antarctic sea ice and the broader Antarctic environment:

    - What have we learned from the satellite data record?
    - What have we learned from available in situ and proxy data records?
    - How can we extend the satellite time series back in time (e.g., using pre-1979 satellite data, ship observations, and other proxies)?
    - What do we need to improve satellite validation of Antarctic sea ice concentration, thickness, and snow cover?
    - What new satellite observations would help improve our understanding?
    - What new in situ and proxy observations would help improve our understanding?

2. Process-based studies of changes in Antarctic sea ice and the broader Antarctic environment:

    - What have we learned from process-based studies?
    - What have we learned about feedbacks in the ice-ocean-atmosphere system and how they may be contributing to the growing sea ice cover?
    - What additional process-based studies could improve our understanding?

3. Models of changes in Antarctic sea ice and the broader Antarctic environment:

    - What have we learned from sea ice and climate models?
    - What are the best projections of future Antarctic sea ice?
    - What types of data synthesis and field experiments are needed for improving sea ice and climate models?
    - What types of focused model intercomparisons could help address model discrepancies?
    - What is needed to improve the forcing fields in Antarctica to improve modeled simulations?

Appendix B
Definitions of Key Terms

Amundsen Sea Low is the climatological area of low pressure located in the South Pacific sector of the Southern Ocean.[1]
Antarctic Circumpolar Current (ACC) is the world's largest ocean current, encircling the Antarctic continent, flowing eastward through the southern portions of the Pacific, Atlantic, and Indian Oceans.[2]

Circumpolar Deep Water (CDW) is the largest water mass by volume transported by the ACC, derived from a mixture of the deep waters from all the world's oceans.[3]
Coupled Model Intercomparison Project (CMIP) is an international activity established in 1995 under the auspices of the World Climate Research Programme (WCRP) and the Working Group on Coupled Modelling (WGCM). CMIP is a standard experimental protocol for studying the output of coupled atmosphere-ocean general circulation models. The comparisons use standardized specifications of model inputs and standardized output formats agreed on by an international committee, and an extensive suite of model outputs is archived and made publicly available for the science and applications communities.[4,5]

Ekman circulation is the net transport of water due to coupling between wind and surface waters, also known as Ekman transport, resulting from the Ekman spiral (how horizontal wind sets surface waters in motion, in which the speed and direction of water motion change with increasing depth).[6]
El Niño–Southern Oscillation (ENSO) is an interannual variability phenomenon that combines the tropical Pacific Ocean surface water temperature over the eastern equatorial Pacific Ocean with air pressure difference across the tropical western Pacific Ocean basin (the Southern Oscillation). ENSO has two coupled phases: El Niño (warm phase) and La Niña (a cool phase).[7]

---

[1] Raphael, M. N., G. J. Marshall, J. Turner, R. L. Fogt, D. Schneider, D. A. Dixon, J. S. Hosking, J. M. Jones, and W. R. Hobbs. 2016. The Amundsen Sea Low: Variability, change, and impact on Antarctic climate. *Bulletin of the American Meteorological Society* 97(1):111-121.
[2] Fyfe, J. C., and O. A. Saenko. 2005. Human-induced change in the Antarctic Circumpolar Current. *Journal of Climate* 18(15):3068-3073.
[3] Dinniman, M. S., J. M. Klinck, and E. E. Hofmann. 2012. Sensitivity of Circumpolar Deep Water transport and ice shelf basal melt along the west Antarctic Peninsula to changes in the winds. *Journal of Climate* 25(14):4799-4816.
[4] NRC (National Research Council). 2012. *National Strategy for Advancing Climate Modeling*. Washington, DC: The National Academies Press.
[5] See http://cmip-pcmdi.llnl.gov.
[6] See http://oceanmotion.org/html/background/ocean-in-motion.htm.
[7] NASEM (National Academies of Sciences, Engineering, and Medicine). 2016. *Frontiers in Decadal Climate Variability: Proceedings of a Workshop*. Washington, DC: The National Academies Press. doi: 10.17226/23552.

Greenhouse Gases (GHGs) are those gaseous constituents of the atmosphere, both natural and anthropogenic, that absorb and emit radiation at specific wavelengths within the spectrum of infrared radiation emitted by the Earth's surface, the atmosphere, and clouds. Water vapor (H_2O), carbon dioxide (CO_2), nitrous oxide (N_2O), methane (CH_4), and ozone (O_3) are the primary greenhouse gases in Earth's atmosphere.[8]

Marginal Ice Zone is a transition zone between the ice edge (often defined by the 15% contour of ice concentration) and the boundary of ice having a concentration greater than 80%.[9]

Ozone Hole is a phenomenon whereby every year, during the Southern Hemisphere spring, a very strong depletion of the ozone layer takes place over the Antarctic region, also caused by human-made chlorine and bromine compounds in combination with the specific meteorological conditions of that region.[10]

Pacific South America (PSA) Pattern is a mode of climate variability in the mid-to-high southern latitudes.[11]
Passive Microwave Sensors are a type of microwave sensor (radiometer) that relies on the thermal emission of microwave signals from objects. The emission is related to the physical temperature and electrical properties of the sensed surface.[12]
Polynyas are areas of persistent open water where one would expect to find sea ice. The water remains open because of processes that prevent sea ice from forming or that quickly move sea ice out of the region.[13]

Reanalysis refers to the reprocessing of observational data spanning a historical period using a consistent modern analysis system.[14]
Representative Concentration Pathways (RCPs) are GHG concentration trajectories used for climate modeling and research.[15]
Rossby Waves are a natural phenomenon in the ocean and atmosphere that owe their properties to Earth's rotation. Oceanic Rossby waves move along the thermocline.[16]

---

[8] IPCC. 2001. Annex II: Glossary of Terms. In *Climate Change 2001: Synthesis Report. A Contribution of Working Groups I, II, and III to the Third Assessment Report of the Integovernmental Panel on Climate Change.* R. T. Watson and the Core Writing Team, eds. Cambridge, UK: Cambridge University Press.
[9] See http://seaiceatlas.snap.uaf.edu/glossary.
[10] IPCC. 2001. Annex II: Glossary of Terms. In *Climate Change 2001: Synthesis Report. A Contribution of Working Groups I, II, and III to the Third Assessment Report of the Integovernmental Panel on Climate Change.* R. T. Watson and the Core Writing Team, eds. Cambridge, UK: Cambridge University Press.
[11] Mo, K. C., and R. W. Higgins. 1998. The Pacific-South American modes and tropical convection during the Southern Hemisphere winter. *Monthly Weather Review* 126(6):1581-1596.
[12] Long, D. G. 2008. Revision. Microwave sensors–active and passive. In *Manual of Remote Sensing, Vol. 3: Remote Sensing for the Earth Sciences.* R. A. Ryerson and A. N. Rencz, eds. Hoboken, NJ: John Wiley and Sons.
[13] See https://nsidc.org/cryosphere/seaice/characteristics/polynyas.html.
[14] NASEM (National Academies of Sciences, Engineering, and Medicine). 2016. *Frontiers in Decadal Climate Variability: Proceedings of a Workshop.* Washington, DC: The National Academies Press. doi: 10.17226/23552.
[15] NRC (National Research Council). 2014. *Linkages Between Arctic Warming and Mid-Latitude Weather Patterns.* Washington, DC: The National Academies Press.

Sea Ice Concentration describes the relative amount of area covered by ice, compared to some reference area, typically reported as a percentage.[17]
Sea Ice Extent is a measurement of the area of ocean where there is at least some sea ice. Usually, scientists define a threshold of minimum concentration to mark the ice edge; the most common cutoff is at 15%.[18]
Sea Ice Persistence is the percent time sea ice concentrations are greater than 15% in a given time period.[19,20]
Sea Ice Volume is a measurement that is estimated through calculations of sea ice thickness with sea ice concentration.[21]
Seasonal Sea Ice is sea ice that melts away and reforms annually. Almost all Southern Ocean or Antarctic sea ice is seasonal sea ice.[22]
Southern Annular Mode (SAM) is the primary mode of atmospheric variability at high southern latitudes.[23]

Westerly Winds are winds from the west toward the east in the mid-latitudes (between 30 and 60 degrees).[24]

---

[16] NASEM (National Academies of Sciences, Engineering, and Medicine). 2016. *Frontiers in Decadal Climate Variability: Proceedings of a Workshop.* Washington, DC: The National Academies Press. doi: 10.17226/23552.
[17] See https://nsidc.org/cryosphere/seaice/data/terminology.html.
[18] See https://nsidc.org/cryosphere/seaice/data/terminology.html.
[19] Finkl, C. W., J. M. Grebmeier, and W. Maslowski. 2015. *The Pacific Arctic Region: Ecosystem Status and Trends in a Rapidly Changing Environment.* Dordrecht: Springer Netherlands.
[20] Stammerjohn, S. E., D. G. Martinson, R. C. Smith, and R. A. Iannuzzi. 2008. Sea ice in the western Antarctic Peninsula region: Spatio-temporal variability from ecological and climate change perspectives. *Deep Sea Research Part II: Topical Studies in Oceanography* 55(18):2041-2058.
[21] See http://www.metoffice.gov.uk/research/areas/cryosphere-oceans/sea-ice/measure.
[22] See https://nsidc.org/cryosphere/sotc/sea_ice.html.
[23] Raphael, M. N., G. J. Marshall, J. Turner, R. L. Fogt, D. Schneider, D. A. Dixon, J. S. Hosking, J. M. Jones, and W. R. Hobbs. 2016. The Amundsen Sea Low: Variability, change, and impact on Antarctic climate. *Bulletin of the American Meteorological Society* 97(1):111-121.
[24] NASEM (National Academies of Sciences, Engineering, and Medicine). 2016. *Frontiers in Decadal Climate Variability: Proceedings of a Workshop.* Washington, DC: The National Academies Press. doi: 10.17226/23552.

Appendix C
Acronyms and Abbreviations

AABW	Antarctic bottom water
ACC	Antarctic Circumpolar Current
ASL	Amundsen Sea Low
CESM	Community Earth System Model
CFC	chlorofluorocarbon
CMIP5	Coupled Model Intercomparison Project Phase 5
CO_2	carbon dioxide
CP	El Niño Modoki
EM	electromagnetic
ENSO	El Niño–Southern Oscillation
Envisat	Environmental satellite
EP	Canonical El Niño
ERA	European reanalysis
ERS	European Remote Sensing
ESMR	Electrically Scanning Microwave Radiometer
GHG	greenhouse gas
ICESat	Ice, Cloud, and land Elevation Satellite
IPCC	Intergovernmental Panel on Climate Change
MIZ	marginal ice zone
MODIS	Moderate Resolution Imaging Spectroradiometer
MPAS	Model for Prediction Across Scales
MSA	methanesulfonic acid
MSLP	mean sea level pressure
NASA	National Aeronautics and Space Administration
NCAR	National Center for Atmospheric Research
PSA	Pacific South American
SAM	Southern Annular Mode
SOSE	Southern Ocean State Estimate
SST	sea surface temperature
UCDW	Upper Circumpolar Deep Water

Appendix D
Workshop Agenda and Participants

January 11-12, 2016
University of Colorado Boulder
CIRES Auditorium-Room 338
216 UCB
Boulder, CO 80309

In contrast to shrinking sea ice in the Arctic, sea ice in the oceans surrounding Antarctica has been more extensive on average in recent years than at any other time in the satellite-observed record. There is not a consensus on the causes of Antarctic sea ice trends or regional variability, and observed patterns are not well represented in Earth system models. The goals of this workshop include to examine the observational record of Antarctic sea ice variability; to further understanding of the processes governing the decadal-scale changes and recent surge in Antarctic sea ice extent; to assess projections of future Antarctic sea ice trends; and to identify ways to improve understanding and modeling of recent and future Antarctic sea ice variability. A summary report will be written based on the workshop's presentations and discussions. This report will not include consensus conclusions or recommendations, but will contain a record of key points emerging from the workshop.

MONDAY, JANUARY 11, 2016

8:15 A.M. Welcome, introductions and purpose of the workshop ***Julienne Stroeve***

SESSION 1 – The Big Picture: Patterns and Drivers of Antarctic Sea Ice Variability

8:45 A.M. BIG PICTURE TALKS *Moderator:* ***Susan Solomon***

Sharon Stammerjohn, University of Colorado Boulder
Cecilia Bitz, University of Washington
Will Hobbs, University of Tasmania

10:05 A.M. Discussion

10:35 A.M. **Break**

SESSION 2 – The Observational Record: Insights and Gaps

Key session questions:

- What does the observational record of Antarctic sea ice extent tell us about sea ice variability and recent trends?

- How do other observations inform our understanding?
- What are the key uncertainties in the observational record?
- What additional observations are critical for improving our understanding of sea ice and its role in the coupled system?

10:50 A.M. OBSERVATIONS PANEL *Moderator:* ***Ted Maksym***

Rob Massom, Australian Antarctic Division
Nathan Kurtz, National Aeronautics and Space Administration-Goddard Space Flight Center
Michael Meredith, British Antarctic Survey
David Bromwich, The Ohio State University
Sarah Das, Woods Hole Oceanographic Institute

11:35 A.M. Opportunity for participants to present 2-minute "speed talks" related to observations[1]

- Comparing Ice Extents from Ice Charts and Passive Microwave, Stephen Ackley, The University of Texas at San Antonio
- Small Scale Sea Ice Kinematics, Ron Kwok, Jet Propulsion Laboratory, California Institute of Technology
- Next Generation European SST Reanalsyis, David Bromwich, The Ohio State University
- The Satellite Passive-Microwave Record of Antarctic Sea Ice Extent, Claire Parkinson, NASA Goddard Space Flight Center

11:50 A.M. Discussion

12:20 P.M. **Working lunch**

1:00 P.M. Breakout groups

Blue Group Leader: Susan Solomon Note-taker: Ali Macalady Rapporteur: Steve Ackley	**Green Group** Leader: Marika Holland Note-taker: Alice Bradley Rapporteur: Sharon Stammerjohn
Red Group Leader: Julienne Stroeve Note-taker: Amanda Staudt Rapporteur: Ron Kwok	**Orange Group** Leader: Xiaojun Yuan Note-taker: Hana Khan Rapporteur: Mike Meredith

1:50 P.M. **Break** (Group leaders, note-takers, and rapporteurs coordinate)

2:10 P.M. Report back from breakout group discussions

---

[1] In addition to scheduled talks by panelists, workshop attendees will have an opportunity to present 2-minute "speed talks" to convey their recent findings or key thoughts to address sessions' organizing questions. Attendees will also participate in panel and breakout group discussions.

3:00 P.M. **Break**

SESSION 3 – Sea Ice Variability in Earth System Models

Key session questions:

- What insights do modeling studies offer regarding the drivers of Antarctic sea ice variability and recent trends?
- What are the current discrepancies between models and observations, and what are possible reasons for such differences?
- What do models portend about the future of sea ice in Antarctica?
- What is needed to improve model representation of sea ice variability?

3:15 P.M. MODELING PANEL *Moderator: **Marika Holland***

Clara Deser, National Center for Atmospheric Research
Lorenzo Polvani, Columbia University
John Fyfe, Environment and Climate Change Canada
Hughes Goosse, Université Catholique de Louvain

3:55 P.M. Opportunity for participants to present 2-minute "speed talks" related to modeling

- Sea Ice Model for Prediction Across Scales (MPAS), Elizabeth Hunke, Los Alamos National Laboratory
- Cloud and Radiation Influence on Southern Ocean Sea Ice, Jennifer Kay, Cooperative Institute for Research in Environmental Sciences, University of Colorado Boulder
- Antarctic Sea Ice Area Trends in the CMIP5 Model, Laura Landrum, National Center for Atmospheric Research
- Using Social Science Research Techniques to Test Communication Strategies for Observed Antarctic Sea Ice Trends, Katy Barnhart, Annenberg Public Policy Center, University of Pennsylvania

4:10 P.M. Discussion

4:40 P.M. Breakout groups

Blue Group Leader: Susan Solomon Note-taker: Ali Macalady Rapporteur: Elizabeth Hunke	**Green Group** Leader: Marilyn Raphael Note-taker: Alice Bradley Rapporteur: Kyle Armour
Red Group Leader: Ted Maksym Note-taker: Amanda Staudt Rapporteur: Alexander Haumann	**Orange Group** Leader: David Holland Note-taker: Hana Khan Rapporteur: David Schneider

5:30 P.M. Adjourn and Reception

TUESDAY, JANUARY 12, 2016

8:30 A.M. Recap of day 1 and plan for day 2 ***Julienne Stroeve***

8:45 A.M. Report back from Session 3 breakout groups

SESSION 4 – Process and Attribution Studies of the Coupled System

Key session questions:

- What is the contribution of tropical ocean-atmosphere variability to Antarctic sea ice variability and trends?
- How does the Southern Ocean and ice shelf melt influence sea ice?
- What is the role of atmospheric phenomena such as the Amundsen Sea Low (ASL)?
- What is the role of anthropogenic forcing (greenhouse gases, ozone)?
- What is the role of natural climate variability?
- What are the key gaps in our understanding of recent sea ice change, and what are the key challenges and opportunities for addressing these gaps?

9:35 A.M. TROPICAL-POLAR CONNECTIONS *Moderator:* ***Xiaojun Yuan***

Graham Simpkins, University of California, Irvine
Xichen Li, Scripps Institution of Oceanography, University of California, San Diego
David Schneider, National Center for Atmospheric Research

10:05 A.M. Discussion

10:25 A.M. OCEAN–SEA ICE AND OCEAN–SHELF ICE–SEA ICE INTERACTIONS *Moderator:* ***David Holland***

Doug Martinson, Lamont-Doherty Earth Observatory
David Ferreira, University of Reading
Kyle Armour, University of Washington

10:55 A.M. Discussion

11:15 A.M. **Break**

11:30 A.M. ATMOSPHERE–SEA ICE INTERACTIONS AND ANTHROPOGENIC FORCING *Moderator:* ***Marilyn Raphael***

Ryan Fogt, The Ohio State University
Alexander Haumann, ETH Zurich
Paul Holland, British Antarctic Survey

12:00 P.M. Opportunity for participants to present 2-minute "speed talks" on processes and attribution

- Southern Ocean Snowfall, Katherine Leonard, Cooperative Institute for Research in Environmental Sciences, University of Colorado Boulder
- Insights from 20th Century Proxies, Will Hobbs, University of Tasmania
- Sea-Ice Induced Ocean Freshening: A Positive Feedback, Alexander Haumann, ETH-Zürich
- Connecting Tropics to Polar Regions, David Bromwich, The Ohio State University
- Zonal Asymmetry in Antarctic Climate Changes, Xichen Li, Scripps Institution of Oceanography
- Impacts of Antarctic Ice Shelf Melting, Ryan Walker, NASA Goddard Space Flight Center
- Ocean Circulation and Topographic Effects on Antarctic Winter Sea Ice and Its Decadal Trends, Lynne Talley, Scripps Institution of Oceanography
- Antarctic Sea Ice Predictability in a Statistical Forecast Model, Xiaojun Yuan, Lamont-Doherty Earth Observatory
- The Seasonality of Antarctic Sea-Ice Trends, Paul Holland, British Antarctic Survey
- Regional Differences in Mechanisms Governing (Seasonal Change in) Ice Extent Around Antarctica, Rob Massom, Australian Antarctic Division
- Polar Ice Trends, Sharon Stammerjohn, Institute of Arctic and Alpine Research (INSTAAR) and University of Colorado Boulder

12:15 P.M. Discussion

12:45 P.M. **Working Lunch**

1:30 P.M. Breakout groups

Blue Group Leader: Susan Solomon Note-taker: Ali Macalady Rapporteur: Will Hobbs	**Green Group** Leader: Marika Holland Note-taker: Alice Bradley Rapporteur: Ryan Walker
Red Group Leader: Julienne Stroeve Note-taker: Amanda Staudt Rapporteur: Cecilia Bitz	**Orange Group** Leader: Xiaojun Yuan Note-taker: Hana Khan Rapporteur: David Bromwich

2:20 P.M. **Break** (Time for group leaders, note-takers, and rapporteurs to coordinate)

2:40 P.M Report back from breakout group discussions

3:30 P.M. Wrap up and final remarks ***Julienne Stroeve***

3:45 P.M. Open workshop adjourns

Workshop on Antarctic Sea Ice Variability in the Southern Ocean-Climate System
Participant List

**Denotes remote participant*

Steve Ackley, The University of Texas at San Antonio
Kyle Armour, University of Washington
Katy Barnhart, University of Pennsylvania
Cecilia Bitz, University of Washington
Scott Borg, National Science Foundation
Alice Bradley, University of Colorado Boulder
David Bromwich, The Ohio State University
John Cassano, University of Colorado Boulder
Sarah Das, Woods Hole Oceanographic Institution
Clara Deser, National Center for Atmospheric Research
David Ferreira, University of Reading
Ryan Fogt, The Ohio State University
William Frey, University Colorado Boulder
John Fyfe, Environment and Climate Change Canada
Hugues Goosse, Université Catholique de Louvain
Alexander Haumann, ETH-Zürich
Will Hobbs, University of Tasmania
David Holland, New York University
Marika Holland, National Center for Atmospheric Research
Paul Holland, British Antarctic Survey
Elizabeth Hunke, Los Alamos National Laboratory
Jen Kay, University of Colorado Boulder
Hana Khan, University of California, Los Angeles
Nathan Kurtz, NASA Goddard Space Flight Center
Ron Kwok, Jet Propulsion Laboratory, California Institute of Technology
Laura Landrum, National Center for Atmospheric Research
Katherine Leonard, University of Colorado Boulder
Xichen Li, Scripps Institution of Oceanography
Alison Macalady, National Academies of Sciences, Engineering, and Medicine
Ted Maksym, Woods Hole Oceanographic Institution
Doug Martinson, Lamont-Doherty Earth Observatory
Rob Massom, Australian Antarctic Division
Mike Meredith, British Antarctic Survey
Caryn Panowicz, National Oceanic and Atmospheric Administration
***Claire Parkinson,** NASA Goddard Space Flight Center
Lorenzo Polvani, Columbia University
Marilyn Raphael, University of California, Los Angeles
Ted Scambos, National Snow and Ice Data Center and University of Colorado Boulder
David Schneider, National Center for Atmospheric Research
Graham Simpkins, University of California, Irvine
Susan Solomon, Massachusetts Institute of Technology
Sharon Stammerjohn, Institute of Arctic and Alpine Research (INSTAAR) and University of Colorado Boulder
Amanda Staudt, National Academies of Sciences, Engineering, and Medicine
Julienne Stroeve, National Snow and Ice Data Center and University of Colorado Boulder

Lynne Talley, Scripps Institution of Oceanography
Ryan Walker, NASA Goddard Space Flight Center
Xiaojun Yuan, Lamont-Doherty Earth Observatory

Appendix E
Invited Speaker Abstracts

Session 1: The Big Picture: Patterns and Drivers of Antarctic Sea Ice Variability

Sharon Stammerjohn
Patterns of Antarctic sea ice variability: Observational insights, gaps, and challenges

One of the most variable but largest seasonal signals on Earth is the six-fold change in Antarctic sea ice extent that occurs every year in the Southern Ocean. Superimposed on this large seasonal transformation is a modest increase in Antarctic sea ice extent over the past four decades. However, this circumpolar average hides a mix of regional and seasonal sea ice extent trends, some of which are large but opposing, while others are modest and fall within the range of natural variability. While sea ice extent estimates are consistent between different satellite retrieval methods, less known or validated are changes in sea ice concentration, thickness, and snow cover. Although we know that regional and seasonal variability is caused by variable exposure to wind and wave events and heat and freshwater inputs, we lack the observations to properly assess their role in driving sea ice changes. Furthermore, differences in regional and seasonal exposure and sensitivity to large-scale climate variability create a variable mix of drivers and feedbacks operating at different time/space scales. Throughout this brief review, key gaps, challenges, and questions will be highlighted to stimulate discussion of strategies for improving assessments of Antarctic sea ice variability.

Cecilia Bitz
Understanding Antarctic climate change: How are we led and misled by models?

Observations of Southern Ocean climate variability indicate a consistent relationship between trends in atmospheric circulation, sea ice extent, and sea surface temperature. Despite an overall poor success by climate models to reproduce the climatology and recent trends in the surface climate, can we use climate models to identify a limited set of leading interactions that might be at play? The basic mechanisms of heat uptake and heat transport in the Southern Ocean that delay global warming are captured in climate models. But there is still vigorous debate about the role of recent wind anomalies, whether they are connected to tropical teleconnections, ozone depletion, or local natural variability. I will discuss why models yield mixed insights into the impacts of winds. I will also consider to what extent models may be simply missing key features, and where effort should be placed toward model improvements, such as coupled interactions with ocean and ice shelves; the influence of waves on sea ice; ocean mixing; the influence of small-scale ocean eddies, katabatic winds, or polynyas; and lastly the proper simulation of clouds and their impact on the surface energy budget.

Will Hobbs
Drivers and forcings of observed Southern Ocean sea ice change

Observations of Southern Ocean sea ice since 1979 show a spatially heterogeneous pattern of trends with a modest overall increase in areal extent, most intense in the austral autumn. By contrast, global coupled-model experiments indicate that the response to anthropogenic forcings is a spatially homogeneous decrease in ice cover, most intense in winter. Does this discrepancy indicate fundamental flaws in the climate models, or the entirely plausible scenario of a modest forced response being overwhelmed by the high internal variability of the system? To answer this question with confidence, it is necessary to consider the physical drivers of change, and to assess whether those drivers are adequately represented in the models. In this presentation I will give a brief overview of atmosphere, ocean, and ice sheet processes that have been hypothesized to contribute to the observed patterns of change, and give a critical assessment of our current level of understanding. I will also outline the current challenges impeding a confident statement regarding the current and future response of Antarctic sea ice in an anthropogenically altered climate.

Session 2: The Observational Record: Insights and Gaps

Rob Massom

Antarctic sea ice observations: Background

This short talk provides an assessment of various datasets available for analyzing large-scale change and variability in Antarctic sea ice, comprising both pack and fast ice and their snow covers, and what they show. In so doing, it highlights key uncertainties, gaps, and issues in the observational record to date. It also underlines the crucial need for improved understanding of complex interactive processes driving the weak overall increase in areal coverage since 1979 that represents the residual of opposing regional (and seasonal) trends. A brief introduction to the modern satellite passive-microwave record and a new coastal fast ice dataset is followed by an assessment of proxy and other data from the pre-1979 era that highlights the importance of placing recent overall and regional trends in longer-term context, and the dependence of trends on the end points chosen. Although widely used, sea ice extent alone is an incomplete/partial descriptor of sea ice state; also required is reliable information on sea ice and snowcover thickness and volume, with consideration of seasonality, different regional settings, and extreme events also being important factors.

Nathan Kurtz

Contribution of Antarctic sea ice thickness observations to understanding of sea ice volume change

Observational data of Antarctic sea ice thickness have been compiled from a wide variety of sources including in situ measurements, undersea sonar, and laser and radar altimetry sources. Each data set provides key information on the overall thickness and volume of Antarctic sea ice; however, there is much that needs to be done to use these records to understand and constrain observed variability and trends. To date, observational sources have shed light on the overall thickness and volume of the Antarctic sea ice pack from a climatological perspective, but uncertainties resulting from spatial sampling, instrument uncertainties, and temporal differences have precluded their use in assessing trends and variability in the volume. However, new advances in observations and promising methods to utilize the results with model records do provide the potential to utilize these records to understand change in the Antarctic sea ice system, which will be the focus of this presentation.

Michael Meredith

Ocean observations and Antarctic sea ice

The ocean is an integral component of the high-latitude air-sea-ice system, and various theories have been advanced concerning the role of the Southern Ocean in recent spatiotemporal changes in Antarctic sea ice. These include, but are not limited to, strengthening of the upper ocean stratification caused by enhanced freshwater inputs, wind-induced acceleration of the northward Ekman transport of cold surface waters, changes in the rate of upwelling of warm waters from the mid-layers of the Southern Ocean and subduction of waters back into the interior, and complex mixed-layer feedbacks that can sustain and promote atmospherically induced sea ice changes. While different models suggest that each of these can be important in different regions and on different time scales, testing the veracity of these models with observational data is challenging. Aside from the surface ocean (which is comparatively well monitored by satellites) and a small number of relatively well-observed sites, the subsurface Southern Ocean is one of the biggest data deserts on the planet, and this is especially true for the ocean beneath the sea ice itself, especially in the winter months when processes critical to sea ice occur. Needed is a much more comprehensive ocean observing including a step-change in the capability for under-ice measurements, as was recently advocated by the Southern Ocean Observing System. This will enable collection of the data required to properly understand the role of the ocean in Antarctic sea ice change and to increase predictive skill concerning how it will change in the future.

David Bromwich

Atmospheric observations over the Antarctic sea ice zone

The Antarctic sea ice zone is almost completely devoid of conventional meteorological observations. Thus, the atmospheric parameters of most relevance to studies of Antarctic sea ice behavior (winds, temperatures, precipitation, downward radiation fluxes, etc.) are most often derived from global atmospheric reanalyses that merge all available observations with short-term model forecasts. For the Southern Ocean, the most important data source is the wide variety of satellite observations. Unfortunately, the discontinuities between satellite missions, which even today emphasize operational needs rather than being of climate quality, lead to jumps and artificial trends in global reanalyses. There are also important shortcomings to the representation of forecast clouds in the atmospheric models used for reanalyses that negatively impact the forecast radiation fluxes. A survey of the performance of contemporary global atmospheric reanalyses for both the modern satellite era (post-1978) and for the centennial time scale will be presented focusing on those variables of most interest to sea ice investigations.

Sarah Das

What can we learn from ice cores about Antarctic sea ice variability and trends prior to the observational era?

Overall Antarctic sea ice coverage appears to be more extensive today than in recent decades, in contrast both to many model predictions as well as to the recent dramatic decline of sea ice in the Arctic. One major challenge to understanding recent changes in Antarctic sea ice, and therefore to also improving predictions of future Antarctic sea ice trends under varying climate scenarios, is the extremely short period of direct observations, which are for the most part limited to the recent decades of the satellite era. And given the high spatial and temporal variability of Antarctic sea ice documented over recent decades, these direct observations may be insufficient to situate recent anomalies and trends into the proper temporal and climatological context. Thus, there is a need to develop new proxy records for reconstructing both longer-term and regionally resolved records of past sea ice conditions around Antarctica, not only for longer time periods (e.g., glacial-interglacial periods), but also critically for more recent centuries to decades. The most likely (and arguably the only possible) candidates for yielding such sea ice proxy information are the depositional records provided by marine sediment cores and ice cores, and indeed this is where many recent and current research efforts have been focused. A number of ice-core chemical constituents originate from aerosols sourced from the sea surface or the marginal sea ice zone, and thus their concentration in ice cores should be strongly influenced by sea ice behavior. The most notable (and best studied) of these to date are sea salts (Na and Cl), which can be sourced both from sea spray as well as the surface of newly formed sea ice, and methanesulfonic acid (MSA), a biogenic marker that is an oxidation product of dimethylsulfide (DMS), a gas characteristic of the near sea-ice zone. Efforts are under way to improve our ability to use variability in ice core–derived Na and MSA concentrations, as well newer glaciochemical proxies in development, to inform us about past sea ice spatial extent and/or its presence or absence through time. Yet many significant challenges remain in understanding the source, transport, and post-depositional behavior of all these marine-derived aerosols, as well as the identification of characteristics of ice sheet environments that would be most suitable for reconstructing past regional sea ice behavior, all of which can be very site specific. In short, moving the utility of glaciochemical records from a qualitative understanding to the quantitative reconstructions needed by the sea ice research community, especially on interannual-to-decadal time scales, remains very much a work in progress. Nonetheless, despite the numerous challenges, ice core records are likely to remain one of the most promising avenues toward the reconstruction of past sea ice variability.

Session 3: Sea Ice Variability in Earth System Models

Clara Deser

Modeling approaches to understanding observed Antarctic sea ice variability and trends

I will provide an overview of the different types of modeling experiments that are used to understand Antarctic sea ice variability and trends and discuss their strengths and limitations. At one end of the spectrum are free-running coupled climate models driven by historical radiative forcing estimates associated with variations in greenhouse gas and stratospheric ozone concentrations as well as anthropogenic and volcanic aerosols. Such simulations show a range of Antarctic sea ice behavior depending on how the model responds to the imposed radiative forcings and its particular chronology of internal (unforced) variability. At the other end of the spectrum are coupled ocean–sea ice models driven by the observed evolution of atmospheric conditions such as winds and air temperatures. Such simulations provide information on how the ocean–ice system responds to the imposed atmospheric conditions, but do not address the origin of the wind and air temperature variations. Intermediate in this hierarchy are coupled models in which sea surface temperatures (SSTs) in a particular region such as the tropical Pacific or North Atlantic are nudged to observations. Such "pacemaker" simulations test the role of remote SST changes on the behavior of Antarctic sea ice via oceanic and atmospheric pathways.

Lorenzo Polvani

Using global climate models to understand recent Antarctic sea ice trends

After several decades of satellite observations, trends in Antarctic sea ice, at a fundamental level, remain poorly understood. General circulation models can be used to shed light on these surprising trends. Notably, such models provide key insights into the role of different anthropogenic forcings. The models also allow us to place the sea ice response caused by individual forcings in the broader context of the natural variability of the Antarctic climate system. Recent findings on these fronts will be reviewed and discussed.

John Fyfe

Have the surface westerlies changed?

Models participating in Phase 5 of the Coupled Model Intercomparison Project (CMIP5) suggest that externally forced strengthening of the Southern Hemisphere surface westerlies may have impacted Antarctic sea ice coverage. However, confirming this in nature is made difficult by the lack of surface wind observations, large uncertainties in estimates from reanalyses, and internal variability. Across reanalyses from 1979 there is no consistent westerly wind trend except in the austral summer, where positive trends of varying magnitude (0.05 to 0.4 m/s per decade) are within the uncertainty range of the CMIP5 simulations. This is suggestive but not confirmation of externally forced westerly wind intensification in observations. Over the shorter period (post-1988) that satellite-based surface wind observations are available, there is a consistent pattern of surface wind change in the South Pacific in observations and reanalyses. However, this pattern of change is dissimilar from the CMIP5 average pattern and is likely a feature of internal variability rather than a signal of externally forced westerly wind intensification. In short, while an externally forced westerly wind impact on Antarctic sea ice coverage is plausible, solid evidence has yet to emerge.

Hugues Goosse

Antarctic sea ice variability and trends: The role of the ocean

The Southern Ocean stores and redistributes heat horizontally and vertically. This can damp or amplify any perturbation, for instance originating in the atmosphere, and provides a strong control on the sea ice variability. The associated mechanisms strongly depend on the mean state of the system; it is thus essential that models reproduce this mean state adequately. Furthermore, a correct initialization of the Southern Ocean is needed for decadal predictions and projections.

Session 4: Process and Attribution Studies of the Coupled System

Graham Simpkins

Tropical Pacific teleconnections to the Southern Hemisphere high latitudes

In this talk, I will review the current understanding of tropical Pacific teleconnections to Antarctica, outlining how interannual and longer-term climate variability are influenced by such processes. On interannual time scales, both eastern Pacific (EP; canonical ENSO) and central Pacific (CP; El Niño-Modoki) SST variability drive pronounced perturbations to the Antarctic surface climate, but with marked seasonal dependence. During the austral cold season (JJAS), both the EP and CP force a similar Rossby wave response, driving anomalous sea ice variability in the West Antarctic region through impacting the Amundsen Sea Low. In the austral warm season (NDJF), by contrast, the EP is linked to zonally symmetric circulation structures that project onto the Southern Annular Mode, whereas only a weak teleconnection is observed from the CP. SST trends, specifically the observed warming in the central/western Pacific, have also been linked to contemporary circulation trends in the Amundsen Sea Low during austral winter and spring. The Rossby wave train mechanism has been proposed to explain such longer-term teleconnections, and as such, West Antarctic surface air temperature and sea ice trends have been further attributed to such processes. There is, however, debate on the extent to which Pacific teleconnections may explain high-latitude trends, and further work is required to clarify controversies in this regard.

Xichen Li

Rossby and Kelvin waves link the Atlantic SST and Antarctic sea ice

During the past three decades, sea surface temperature (SST) over the north and tropical Atlantic exhibited a significant warming trend. Observational data and model simulations link this SST anomaly to the climate changes around Antarctica. These teleconnections critically depend on stationary Rossby waves and are thus sensitive to the background flow, particularly the subtropical and midlatitude jets. Using a hierarchy of climate models, we show that these jets reflect, guide, and focus the Rossby wave trains on western Antarctica, modifying the regional atmospheric circulation, and thus the Antarctic sea ice distribution. Further analyses show that the recently observed SST trends in different tropical ocean basins are linked with each other through atmospheric Rossby and Kelvin waves. These dynamics build multiple pathways from the tropical ocean basins to Antarctica, and thus further complicate the mechanisms of the Antarctic climate variability.

David Schneider

The role of tropical teleconnections in recent Antarctic climate change

This talk will review the role of tropical teleconnections in driving recent trends in the atmospheric circulation around Antarctica and discuss how this is important for understanding the trends in Antarctic sea ice extent. In particular, a series of transient atmospheric model experiments is evaluated to show that changes in tropical sea surface temperatures have played the dominant role in the deepening of the Amundsen Sea Low during the satellite era (which coincides with the modern sea ice record). The impacts of stratospheric ozone depletion on the strengthening of the Southern Ocean westerlies are most evident with a time horizon extending back to about 1960. This knowledge of the drivers of recent circulation changes informs a new set of coupled model experiments in which the tropical Pacific sea surface temperatures are relaxed to the observed values, thereby constraining the model to the observed evolution of tropical climate (including the El Niño–Southern Oscillation and the Pacific Decadal Oscillation phenomena). While these simulations show improvements in simulating the recent atmospheric circulation changes over the Southern Ocean compared to their unconstrained counterparts, they also illustrate the need for improving the representation of physical processes contributing to the evolution of sea ice in coupled models.

Doug Martinson

Antarctic ocean–ice interactions

This talk focuses on the unique ocean–sea ice interactions whereby brine formation during sea ice growth drives upper ocean convection into the permanent pycnocline that releases heat into the mixed layer melting, or preventing from forming a large amount of sea ice. This keeps the sea ice thin (~60 cm, also making it easy to melt in the spring forcing a seasonal sea ice cover). Specifically, for every one unit of ice, the entrainment heat flux releases enough heat to melt 5-20 units of ice in both the open ocean Weddell gyre and continental shelf west Antarctic Peninsula regions. The presence of the cold halocline layer (CHL) in the Arctic prevents this feedback allowing that ice to grow extremely thick (meters) not as easy to melt completely in summer.

The Antarctic feedback is the result of a predominantly vertical system (halocline and thermocline coincide with depth) with a very warm deep layer (Upper Circumpolar Deep Water, UCDW) underlying the pycnocline. The interactions are well predicted from an analytic vertical model whose scaling laws using easily observed variables allow excellent estimates of the maximum ice thickness, magnitude of the feedback, ocean heat flux, and other characteristics of the temporal evolution of the system. UCDW has been warming dramatically the past several decades, reflecting the warming of global deep water. This is the water that melts the underside of the ice streams draining the west Antarctic Ice sheet into the Amundsen Sea Embayment (greatly accelerating their drainage into the ocean, raising global sea level). The warming ensures strong melt, even if the circulation beneath the ice margins slows down.

Overall slight expansion of the sea ice extent around Antarctica, the consequence of expansion in some areas and contraction in others, is not understood yet, but the location of the Antarctic Circumpolar Current (ACC) limits the areal extent of polar water capable of forming ice, and the warming of the surface waters across the ACC limits the distance sea ice can be expanded to the north. So it is *speculated* here that subtle shifts in the ACC via changes in the westerlies may actually account for the statistical zonal sea ice expansion.

David Ferreira

The two–time scale response of Antarctic sea surface temperature and sea ice cover to ozone depletion

In recent modeling work, we showed that the response of Antarctic sea surface temperature and sea ice cover to abrupt ozone depletion has two phases: a fast interannual (~1-5 y) adjustment in which the surface ocean cools and sea ice cover increases, followed by a slower decadal trend leading to a warming of the surface ocean and a reduction of sea ice cover. This result reconciles diverging views, found in the literature, on the relationship between ozone depletion and Antarctic sea ice changes. I will briefly describe the dynamics behind the two–time scale response and the sources of uncertainties in the modeled response. I will then discuss implications of the two–time scale response to evaluate the contribution of ozone depletion (and recovery) to the observed decadal trends and future trends of the Antarctic sea ice.

Kyle Armour

The ocean's role in Southern Ocean climate change: Warming delayed by circumpolar upwelling, and cooling driven by trends in SAM

The region of ocean surface south of the Antarctic Circumpolar Current (ACC) has warmed much more slowly than the global ocean since the 1950s, and has cooled since about 1980—coincident with an overall expansion of the Southern Ocean sea ice cover. Here we argue that ocean circulation plays a critical role in these observed changes.

Comprehensive (CMIP5) and idealized climate model simulations robustly simulate delayed sea surface warming around Antarctica in response to greenhouse gas forcing. Using models and observations, we show that the primary driver of this warming delay is the Southern Ocean's background meridional overturning circulation, which upwells unmodified water to the surface from depth and advects surface waters northward. It is thus against a background of gradual, greenhouse gas–induced warming—rather than the rapid warming observed in the Arctic—that multidecadal Southern Ocean temperature and sea ice trends must be understood.

One proposed driver of the observed cooling and sea ice expansion in recent decades is the trend in the large-scale westerly winds, often characterized as a trend toward a positive phase of the Southern Annular Mode (SAM). Yet, targeted SAM perturbation experiments within comprehensive climate models generally find enhanced warming and sea ice loss. We summarize recent analyses of the Southern Ocean's response to SAM changes within the preindustrial control simulations of CMIP5 models. Those models that simulate a Southern Ocean climatology that most closely matches observations robustly show cooling and sea ice expansion in response to trends in SAM. We argue that the CMIP5 models have been unable to replicate observed Southern Ocean changes due to a confluence of biases in their ocean climatologies and SAM trends that are too weak compared to observations.

Ryan Fogt

The role of anthropogenic forcing and natural variability on Southern Hemisphere atmospheric circulation trends

This presentation will provide a brief overview of the roles of natural variability in comparison to anthropogenic forcing on atmospheric circulation trends near Antarctica, with direct implications for changes in sea ice parameters (season length, concentration, and extent). The primary focus will be on the Amundsen Sea Low (ASL), where the strongest regional sea ice changes have been observed. Notably, studies show the strongest influence from ozone depletion (leading) and greenhouse gases (a smaller contribution) in austral summer, both leading to a deepening of the ASL. Forced atmospheric circulation changes in other seasons are more model dependent, with results varying across studies. Tropical variability alone appears sufficient to capture a large portion of the variability (and even recent changes) in the non-summer ASL, and a recent study even suggests that ongoing sea ice trends are within the bounds of natural, unforced climate variability. Together these studies all suggest there is still some debate on the forcing on recent regional atmospheric circulation changes, and that in particular features of the ASL must be adequately simulated in order to accurately depict regional sea ice trends.

Alex Haumann

Dependence of Antarctic sea ice trends on zonally asymmetric atmospheric circulation changes and ocean stratification

In this presentation, I will illustrate how zonal asymmetries in the large-scale circulation around Antarctica drive variations and long-term trends in Antarctic sea ice and argue that this process is not fully captured by coarse-resolution climate models. The observed large-scale circulation pattern is forcing an outflow of cold air from the two large ice-shelf regions on to the sea ice in the Ross and Weddell Seas. The sea ice strongly responds to the changes of this cold-air outflow with an increased northward transport of sea ice and an expansion of the ice edge when more cold air is leaving the continent, as observed over recent decades. Simulations with a global climate model show that this process is largely absent in the model due to rather zonally symmetric large-scale circulation changes in response to the anthropogenic forcing. I will argue that the underestimated zonal asymmetries in the atmospheric circulation are also causing a much weaker ocean surface stratification in the model. This results in a much higher sensitivity of the sea ice to changes in the subsurface heat fluxes, explaining why simulated sea ice changes are driven by a different mechanism than the changes observed in recent decades.

Paul Holland

Seasonality of Antarctic sea ice trends

The Antarctic sea ice increase features opposing regional trends with a strong seasonal signal. These spatial and temporal patterns can offer significant insight. I will show trends in ice "intensification" (the rate of change of ice concentration), which can be viewed as trends in ice growth/melting or dynamics. The results suggest that the Antarctic sea ice trends originate in spring. The largest concentration trends, in autumn, appear to be caused by trends in intensification during spring. Autumn intensification trends directly oppose autumn concentration trends in most places, seemingly as a result of ice and ocean feedbacks. This causes a problem with the linkage between wind trends and ice trends, which agree in autumn, but do not agree in spring. This suggests that we need to understand more about how autumn trends affect the winter ice cover and spring melting.

Appendix F
Biographical Sketches of Planning Committee Members

Dr. Julienne Stroeve, Chair
National Snow and Ice Data Center
Dr. Julienne Stroeve is a Senior Research Scientist at the National Snow and Ice Data Center (NSIDC) within the Cooperative Institute for Research in Environmental Sciences (CIRES). She received a Ph.D. in geography from the University of Colorado Boulder for her work in understanding Greenland climate variability. Subsequently her Arctic research interests have focused on the sea ice cover and include sea ice predictability, climate change, and associated local and large-scale impacts. She has conducted several Arctic field campaigns. Dr. Stroeve's work has been featured in numerous magazines, news reports, radio shows, and television documentaries. She has given keynote addresses around the world on Arctic climate issues and briefed former Vice President Al Gore. Dr. Stroeve has published more than 50 articles in peer-reviewed journals and contributed to several national and international reports on Arctic climate change.

Dr. David Holland
New York University
Dr. David Holland is a professor of mathematics in the Center for Atmosphere Ocean Science at New York University. His research interests include geophysical fluid dynamics, computational fluid dynamics, climate system dynamics, and ice-ocean interaction. He has published more than 60 peer-reviewed articles on polar environmental science, many regarding tidal influence on the melting and freezing of glaciers and ice shelves. Dr. Holland received his Ph.D. in atmosphere-ocean science from McGill University in 1993.

Dr. Marika Holland
National Center for Atmospheric Research
Dr. Marika Holland is a Senior Scientist in the Climate and Global Dynamics Laboratory at the National Center for Atmospheric Research (NCAR). Her research interests are focused on the role of sea ice in the climate system, including secular sea ice change, sea ice predictability, and polar climate variability. Dr. Holland has extensive experience using coupled climate models to study polar climate variability and change, and has been active in the development of improved sea ice models for climate simulations. She has served as co-chair for the Polar Climate Working Group of the Community Earth System Model, and Chief Scientist for the Community Earth System Model project. She has been a contributing author on the Intergovernmental Panel on Climate Change third, fourth, and fifth assessment reports, contributed to numerous other national and international assessments on the changing Arctic climate, and is an author on more than 80 peer-reviewed publications. Dr. Holland received her Ph.D. in atmosphere and ocean sciences from the University of Colorado in 1997 and held a postdoctoral fellowship at the University of Victoria in British Columbia before joining the scientific staff of NCAR in 1999.

Dr. Ted Maksym
Woods Hole Oceanographic Institution
Dr. Ted Maksym is an Assistant Scientist at the Woods Hole Oceanographic Institution. His research interests include processes driving seasonal sea ice formation and decay, ice-ocean-atmosphere interactions, and distribution of sea ice thickness and its snow cover. A particular focus is the use of

autonomous platforms for observing the ice cover, including the use of autonomous underwater vehicles (AUVs) to assess the state of the ice cover. He is a veteran of 12 research expeditions to the Antarctic and Arctic, including leading the British IceBell expedition which was the first to coordinate aircraft and UAV observations of ice thickness distribution in the Antarctic. He received his Ph.D. from the University of Alaska in 2001.

Dr. Marilyn Raphael
University of California, Los Angeles
Dr. Marilyn Raphael is a professor and former Chair of the Department of Geography at the University of California, Los Angeles. She is also co-chair of the SCAR expert group Antarctic Sea ice Processes and Climate (ASPeCt) and co-lead of the WCRP's Polar Climate Predictability Initiative (PCPI). Her research interests are in climate variability and change particularly in the high-latitude Southern Hemisphere. Her research focuses on understanding the interaction between Antarctic sea ice variability and the large-scale atmospheric circulation using both large-scale observational databases and global climate models. Her current projects focus on understanding the recent trends in the west Antarctic sea ice zone with an emphasis on improving the simulation of sea ice and the atmosphere in the Southern Hemisphere. Dr. Raphael also does work on the Santa Ana Winds of California. Dr. Raphael has served on two NRC committees: Future Outlook for Antarctic Science and Carbon Dioxide Stabilization. Dr. Raphael received her Ph.D. in geography from The Ohio State University.

Dr. Susan Solomon
Massachusetts Institute of Technology
Dr. Susan Solomon is internationally recognized as a leader in atmospheric science, particularly for her insights in explaining the cause of the Antarctic ozone "hole." She and her colleagues have made important contributions to understanding chemistry/climate coupling, including leading research on the irreversibility of global warming linked to anthropogenic carbon dioxide emissions, and on the influence of the ozone hole on the climate of the Southern Hemisphere. Her current focus is on issues relating to both atmospheric chemistry and climate change. Dr. Solomon received her Ph.D. in chemistry from the University of California, Berkeley, in 1981, and served as Senior Scientist at the Aeronomy Laboratory of the National Oceanic and Atmospheric Administration before becoming a professor of atmospheric chemistry and climate science at the Massachusetts Institute of Technology in 2012.

Dr. Xiaojun Yuan
Lamont-Doherty Earth Observatory of Columbia University
Dr. Xiaojun Yuan received a bachelor's degree in physical oceanography from Shangdon College of Oceanography, China, in 1982. After spending 4 years at the National Research Center for Environmental Forecast in Beijing, she came to the United States and worked with the Scripps Institution of Oceanography studying the persistence of sea surface temperature anomalies in the North Pacific and its influence on the climate variability of the Northern Hemisphere. She then enrolled in the Ph.D. program at Scripps under the supervision of Professor Talley, where she investigated oceanic fronts in the North Pacific using historical conductivity-temperature-depth measurements together with surface fluxes. Upon receiving her Ph.D. in 1994, she moved to New York and started postdoctoral studies with Dr. Douglas Martinson at Lamont-Doherty Earth Observatory of Columbia University, where she then focused on Antarctic sea ice and its relationships with global climate. She was then promoted to Doherty Associate Research Scientist in 1998, to Doherty Research Scientist in 2004, and to Lamont Research Professor in 2013.